T0323814

Green Internet of Things for Smart Cities

Green Internet of Things for Smart Cities

Concepts, Implications, and Challenges

Surjeet Dalal, Vivek Jaglan, and Dac-Nhuong Le

CRC Press
Taylor & Francis Group
Boca Raton London New York

CRC Press is an imprint of the
Taylor & Francis Group, an **informa** business

First edition published 2021
by CRC Press
6000 Broken Sound Parkway NW, Suite 300, Boca Raton, FL 33487-2742

and by CRC Press
2 Park Square, Milton Park, Abingdon, Oxon OX14 4RN

© 2021 Taylor & Francis Group, LLC

CRC Press is an imprint of Taylor & Francis Group, LLC

Library of Congress Cataloging-in-Publication Data
Names: Dalal, Surjeet, editor. | Jaglan, Vivek, editor. | Le, Dac-Nhuong, 1983– editor.
Title: Green internet of things for smart cities: concepts, implications,
and challenges / edited by Surjeet Dalal, Vivek Jaglan, and Dac-Nhuong Le.
Description: First edition. | Boca Raton, FL : CRC Press, 2021. |
Series: Green engineering and technology : concepts and applications |
Includes bibliographical references and index.
Identifiers: LCCN 2020042858 (print) | LCCN 2020042859 (ebook) |
ISBN 9780367858537 (hardback) | ISBN 9781003032397 (ebook)
Subjects: LCSH: Smart cities. | Internet of things–Energy conservation. |
Sustainable engineering.
Classification: LCC TD159.4 .G75 2021 (print) | LCC TD159.4 (ebook) |
DDC 307.760285/4678–dc23
LC record available at https://lccn.loc.gov/2020042858
LC ebook record available at https://lccn.loc.gov/2020042859

ISBN: 978-0-367-85853-7 (hbk)
ISBN: 978-0-367-71271-6 (pbk)
ISBN: 978-1-003-03239-7 (ebk)

Typeset in Times
by Newgen Publishing UK

Contents

Preface

Internet of Things (IoT) alludes to physical and virtual items that have special behaviors and are associated with the web to give smart applications that make vitality, coordination, modern control, retail, horticulture, and numerous different spaces "more brilliant." Internet of Things is another insurgency of the Internet that is quickly assembling force driven by the progressions in sensor systems, cell phones, remote interchanges, systems administration, and cloud innovations. Specialists conjecture that in 2050 there will be a sum of 50 billion gadgets/things associated with the web. This book discusses the Internet of Things for smart cities and making the lifestyle of people secure and comfortable. The run of the mill beginner is relied upon to have finished a few courses in programming utilizing conventional significant level dialects at the school level and is either a senior or a starting alumni understudy in one of the science, technology, engineering, and maths (STEM) fields. Like our companion book on *Cloud Computing*, we have attempted to compose a complete book that gives information through a vivid "hands-on" approach, where the beginner is given the important direction and information to design IoT applications used in smart cities and industry levels. Simultaneous improvement of reasonable applications that goes with customary instructional material inside the book further upgrades the learning procedure. The website for this book contains extra help for guidance and learning.

The tremendous technology development in the twenty-first century has many advantages. However, the growth of technology has high energy demands accompanied with e-waste and hazardous emissions. In this book, we survey and identify the most critical technologies used for green IoT and keeping our environment and society smarter and green. The ICT revolution (i.e., FRID, WSN, M2M, communication network, Internet, DC, and CC) has qualitatively augmented the capability for greening IoT. Based on the critical factors of ICT technologies, the things around us will become smarter to perform specific tasks autonomously, rendering a new type of green communication between humans and things and also among things themselves, where bandwidth utilization is maximized and hazardous emission mitigated, and power consumption is reduced optimally. Future suggestions have been touched upon for efficiently and effectively improving green IoT-based applications. This book provides effective insight for anyone who wishes to research the field of green IoT. The trends and prospective future of green IoT are provided.

The bright future of green IoT will change tomorrow's environment to become healthier and green, very high QoS, socially and environmentally sustainable and economically also. Nowadays, the most exciting areas focus on greening things such as green communication and networking, green design and implementations, green IoT services and applications, energy-saving strategies, integrated RFIDs and sensor networks, mobility, and network management, the cooperation of homogeneous and heterogeneous networks, smart objects, and green localization.

Editors

Surjeet Dalal received his Ph.D. degree in 2014 from Suresh Gyan Vihar University Jaipur (Rajasthan) and M. Tech degree in 2010 from PDM College of Engineering, Bahadurgarh Haryana. He has completed B. Tech (Computer Science & Engineering) from Jind institute of Engineering & Technology Jind (Haryana) in 2005. He is working as Associate Professor in CSE department in SRM University, Sonepat, Haryana. His current research arca is artificial intelligence, cloud computing and IoT. He has presented more than 20 papers in the national/international conferences. He has guided Ph.D. scholars and M. Tech. students in different research areas. He has published one book on cloud computing and registered a patent recently. He is the reviewer of many national/international journals of repute in India and abroad. He is a professional member of various professional and research committees. He is a professional member of CSI India, IEEE New York, IETE Chandigarh and ISTE-AICTE New Delhi.

Vivek Jaglan received his Ph.D. degree in 2012 from Suresh Gyan Vihar University Jaipur (Rajasthan) and M. Tech degree from C.D.L.U. Campus Haryana. He has completed B. Tech (Computer Science and Engineering) from Hindu College of Engineering, Industrial Area Sonepat. He is working as Associate Professor in CSE department in Amity University, Gurugram, Haryana. His current research area is artificial intelligence, neural networks and fuzzy logic, and IoT. He has presented more than 40 papers in national/international conferences.

He has published more than 30 papers in the national and international journals. He has published one book on cloud computing and registered a patent recently

Dac-Nhuong Le has a M.Sc. and Ph.D. in computer science from Vietnam National University, Vietnam, in 2009 and 2015, respectively. He is Deputy-Head of Faculty of Information Technology, Haiphong University, Vietnam.

He has a total academic teaching experience of 12 years with more than 50 publications in reputed international conferences, journals and online book chapter contributions (Indexed By: SCI, SCIE, SSCI, ESCI, Scopus, ACM, DBLP). His areas of research include: evaluation computing and approximate algorithms, network communication, security and vulnerability, network performance analysis and simulation, cloud computing, IoT and image processing in biomedical.

His core work in network security, soft computing and IoT and image processing in biomedical. Recently, he has been the technique program committee, the technique reviews, the track chair for international conferences: FICTA 2014, CSI 2014, IC4SD 2015, ICICT 2015, INDIA 2015, IC3T 2015, INDIA 2016, FICTA 2016, ICDECT 2016, IUKM 2016, INDIA 2017, CISC 2017, FICTA 2018, ICICC 2018, CITAM 2018 under Springer-ASIC/LNAI Series. Presently, he is serving in the editorial board of international journals and he has authored eight computer science books by Springer, Wiley, Chapman and Hall/CRC Press, Lambert Publication, Scholar Press, and VSRD Academic Publishing.

Contributors

Abhay Agarwal
Panipat Institute of Engineering &
    Technology
Samalkha, Haryana, India

Prerna Agarwal
Assistant Professor, JIMSTEC
Greater Noida, Uttar Pradesh, India

Akshat Agrawal
Amity University
Manesar, Haryana, India

Shakti Arora
Panipat Institute of Engineering &
    Technology
Samalkha, Haryana, India

Vijay Anant Athavale
Panipat Institute of Engineering &
    Technology, Samalkha
Haryana, India

Vikram Bali
JSS Academy of Technical Education
Greater Noida, Uttar Pradesh, India

Reenu Batra
SGT University
Gurugram, Haryana, India

Jyoti Chauhan
Department of CSE, SRM University
Delhi-NCR, Sonipat, Haryana, India

Nidhi Chawla
Department of CSE, SRM University
Delhi-NCR, Sonipat, Haryana, India

Neeraj Dahiya
Department of CSE, SRM University
Delhi-NCR, Sonipat, Haryana, India

Sandeep Dalal
Assistant Professor, Maharshi Dayanand
    University
Rohtak, Haryana, India

Surjeet Dalal
Department of CSE, SRM University,
    Delhi-NCR
Sonipat, Haryana, India

Shivashish Dhondiyal
Graphic Era Hill University
Dehradun, Uttarakhand, India

Ashutosh Dixit
J. C. Bose University of Science and
    Technology, YMCA
Faridabad, India

Puneet Garg
JCBUST YMCA, Faridabad
Haryana, India

Umang Garg
Graphic Era Hill University,
Dehradun, Uttarakhand, India

Amit Kumar Goel
Galgotia University,
Greater Noida, Uttar Pradesh, India

Mayank Goel
Department of Systemics, University of
    Petroleum & Energy Studies
Dehradun, Uttar Pradesh, India

Puneet Goswami
Department of CSE, SRM University
Delhi-NCR, Sonipat, Haryana, India

Aastha Gour
Graphic Era Hill University
Dehradun, Uttar Pradesh, India

Neha Gupta
Graphic Era Hill University
Dehradun, Uttarakhand, India

Shipra Gupta
Graphic Era Hill University
Dehradun, Uttarakhand, India

Amber Hayat
University of Petroleum & Energy
  Studies
Dehradun. Utarkhand, India

Akhtar Husain
Department of Computer Science & IT,
  MJP Rohilkhand University
Bareilly, Uttar Pradesh, India

J. S. Kalra
Graphic Era Hill University
Dehradun, Uttarakhand, India

D. Kapil
Graphic Era Hill University
Dehradun, Uttarakhand, India

Ajay Kumar
Department of CSE, JSS Academy of
  Technical Education
Greater Noida, Uttar Pradesh, India

Vijay Kumar
Graphic Era Hill University
Dehradun, Uttarakhand, India

Dac-Nhuong Le
Faculty of Information Technology,
Haiphong University, Vietnam

N. Mehra
Graphic Era Hill University
Dehradun, Uttarakhand, India

Anand Mohan
L.N. Mithila University, Darbhanga,
  Bihar, India

Jyoti Nanwal
J.C. Bose University of Science and
  Technology, YMCA
Faridabad, Haryana, India

Nhu Gia Nguyen
Duy Tan University
Danang, Vietnam

Rajesh Pant
Graphic Era Hill University
Dehradun, Uttrakhand, India

Shrivastava Pranav
GLBITM
Greater Noida, Uttar Pradesh, India

N. Punera
Graphic Era Hill University
Dehradun, Uttarakhand, India

Bijeta Seth
SRM University
Delhi-NCR, Sonipat, Haryana, India

Preeti Sethi
J. C. Bose University of Science and
  Technology, YMCA
Faridabad, Haryana, India

Shalini Shori
Scholar, DAVV
Indore, Madhya Pradesh, India

Prachi Shori
Graphic Era Hill University
Dehradun, Uttarakhand, India

Virendra Kumar Shrivastava
SGT University
Gurugram, Haryana, India

Shalini Singh
Graphic Era Hill University, Dehradun,
  Uttarakhand, India

Kamna Solanki
Assistant Professor, Maharshi Dayanand
  University, Rohtak, Haryana, India

Parveen Poon Terang
JSS Academy of Technical Education
Greater Noida, Uttar Pradesh, India

Meenu Vijarania
Department of CSE, Amity University
Manesar, Haryana, India

Vikas Yadav
Department of CSE, Amity University
  Manesar, Haryana, India

1 Green-IoT (G-IoT)
Technological Need for Sustainable Development and Smart World

Sandeep Dalal and Kamna Solanki

CONTENTS

1.1 INTRODUCTION

The term 'Internet of Things' was first coined by Kevin Ashton as the title of a PowerPoint presentation. Kevin had shared an idea of using RFID (Radio Frequency Identification) chips on consumer goods for the automatic tracking of stock levels in storehouses. Afterwards, IoT became an emerging concept and revolutionized the world by connecting billions of devices together. The IoT devices work by sensing, collecting, and transmitting the vital information gathered from the surroundings as depicted in Figure 1.1 and Figure 1.2. However, massive energy is consumed in

1

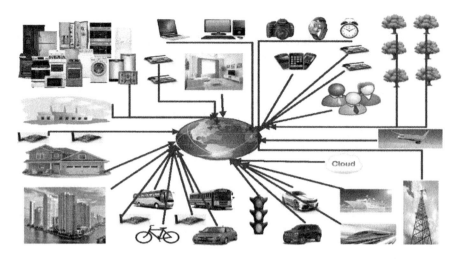

FIGURE 1.1 An example of IoT

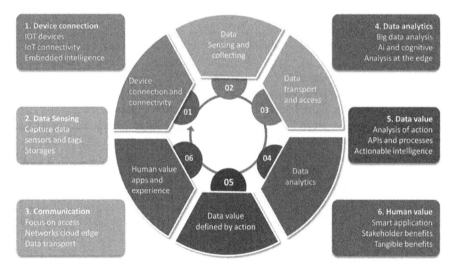

FIGURE 1.2 IoT process cycle

exchanging and sharing information amongst billions of devices, along with emission of radiation and greenhouse gases by these IoT devices.

Massive progress in usage of smartphones and tablet PCs took the number of Internet-connected devices to 12.5 billion in 2010, while the population of the world rose to 6.8 billion, making the "number of connected devices per person" more than 1 (1.84 to be precise) for the first time in history. Cisco IBSG estimated that there would be 25 billion Internet-connected devices by 2015 and 50 billion Internet connected devices by 2020 as shown in Figure 1.3. It is important to note that these figures did

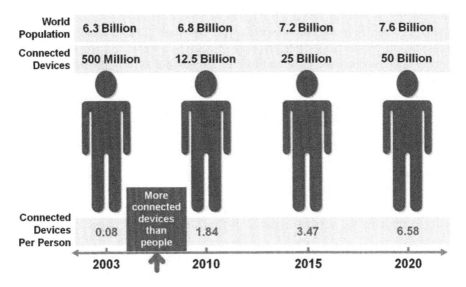

FIGURE 1.3   Predicted increase of internet connected devices by Cisco

not take rapid developments in Internet or computer technology into account; the statistics reported were based on the growth rate and situation at that time. A more recent forecast is from GSMA Intelligence in June 2018, which predicts an estimated 25 billion devices in 2025. "This is consistent with Gartner's 2017 estimate of 20 billion by 2020. We have seen a decline in 2020 estimates from 50 billion to 20 billion over the past five years. This is good reminder to be skeptical about over-enthusiasm early in a technology's lifecycle" [2].

IoT has also brought some obstacles to the environment. For the natural environment, the most important issues are increasing the volume of e-waste, energy consumption, and CO_2 emissions.

1.1.1   Huge Electronic Waste

Nowadays, the whole world is facing an e-waste problem. With the thousands of IoT devices flooding the market in the coming years, non-compatible devices may even end up in landfill. The number is also expected to rise at a faster pace with the speed of progress of the IoT. The problem may not be evident now, but that is something we need to remember for sustainable development. By the end of 2025, the world's population will generate 2.2 billion tons of waste annually. So, global spending on the e-waste management will reach $375 billion, but it's not all about money.

1.1.2   Energy Consumption

The second challenge is its use of electricity. IoT networks require a huge set of data for processing and meeting their needs. As a result, the data centers' energy consumption is also vast and extensive. The materials needed to produce the energy add a huge burden to the atmosphere as well. Although big data centers are trying to use as little power as possible, the energy sector as a whole will still be affected. Energy and resources are also used to create thousands of new products, another source of energy demand generated by the IoT.

1.1.3 CO_2 EMISSIONS, RADIATION EXPOSURE, AND GREENHOUSE EFFECT

A third major problem is the release of CO_2 and other greenhouse gases from IoT devices. The greenhouse effect of these gases will cause environmental imbalance. Several studies on green IoT have suggested approaches which produce significant reductions in CO_2 emissions, cost savings, and low radiation exposure.

Features such as system robustness and energy efficiency are critical. Green IoT seeks to minimize the negative impact of IoT on human and environment.

1.2   GREEN IOT

Green usually means "planning and investing in a technology network that serves the needs of today as well as today's needs while conserving energy and saving money". The Green Internet of Things focuses primarily on the energy efficiency and the reliability of the IoT concepts.

> Green IoT is defined as the energy-efficient way in IoT to either reduce or eradicate the green- house effect caused by existing applications. Green IoT is the process of efficiently producing, developing, disposing of computers, servers, using and associating subsystems (i.e., printers, displays, communications equipment and storage devices), with a reduced negative impact on society and the environment.
>
> [4]

Going for green IoT requires new resources to reduce the negative impact of IoT on human health and climate disruption. Greening IoT's primary goal is to reduce CO_2 emissions and waste, harness environmental conservation, and mitigate the costs of operating stuff and power consumption. Reduction of the energy consumption of IoT devices is required to make the world healthier. Green IoT offers a high potential to support economic growth and environmental sustainability through the advancement of greening ICT technologies.

Major objectives of Green IoT are:

1. Energy efficient procedures adopted by IoT
2. Reduce greenhouse effects of existing IoT applications
3. Reduce energy consumption and CO_2 emissions
4. Electronic waste management
5. Usage of surrounding environment to assist in generating power supply

FIGURE 1.4 Focus of Green IoT

1.2.1 FOCUS OF GREEN IOT

Green IoT can be defined as "The study and practice of designing, using, manufacturing, and disposing of servers, computers, and associated subsystems such as monitors, storage devices, printers, and communication network systems efficiently and effectively with minimal or no impact on the environment" [4].

Green IoT thus contributes to protection of natural resources, minimizing the impact of technology on the atmosphere and human health, and significantly reducing costs. Therefore, green IoT really focuses on the following points (see Figure 1.4):

- Green Design: Design energy-efficient modules, computers, and servers and cooling equipment for green IoT components.
- Green Production: Development of electronic components, computers, and other related subsystems with no negative impact on the environment.
- Green Utilization: Mitigating the energy consumption of IoT devices, computers, and other related systems, and using them in an environment-friendly way.
- Green Disposal: Refurbishing, recycling, and reusing the IoT devices to minimize resource wastage and e-waste production.

1.2.2 GREEN IOT TECHNOLOGIES AND GREEN ENVIRONMENT

Green IoT revolves around the cross-cutting technologies Design Technologies, Leverage Technologies, and Enabling Technologies as shown in Figure 1.5.

- Design Technologies apply to energy efficiency of systems, communications protocols, network architectures, and interconnections.
- Leverage Technologies apply to reducing carbon emissions and increasing energy efficiency. It is also known as green ICT (Information and Communication Technologies).
- Enabling technologies include various technologies such as green RFID, sensor networks, cellular networks, machine-to-machine networking, communications and energy harvesting tools, intelligent radio, cloud computing, and big data

FIGURE 1.5   Green IoT technologies

analysis. With these enabling technologies, green IoT offers great potential to boost economic and environmental performance.

The NIC (National Intelligence Council, India) encourages research and development to save resources when connecting to the real world using the G-IoT tools. Considering energy efficiency, the green energy IoT definition can be defined as

The energy efficient procedures (hardware or software) adopted by IoT either to promote the reduction of the greenhouse effect of existing applications and services or to reduce the impact of the greenhouse effect of IoT itself. In the earlier case, the use of IoT would help to reduce the greenhouse effect, whereas in the latter case consideration will be taken to further maximize the IoT greenhouse footprint. The entire life cycle of green IoT should focus on green design, green production, green utilization and finally green disposal/recycling to have no or very small impact on the environment.

[7]

The Green IoT environment uses smart technologies everywhere to create a sustainable environment (Figure 1.6).

With reduction in energy consumption and harmful gas emissions, with resource investment and reducing pollution, IoT becomes more efficient thanks to green IoT technologies. Green IoT plays an important role in deploying IoT to achieve a smart and sustainable world, improving environmental conservation and minimizing power consumption. Green IoT assists in creating smart homes, smart things, smart cities, smart devices, and a smart world.

1.2.3   Green IoT Implementation/Communication Components

Green IoT may be implemented by using green internet technology, RFIDs, data centers, wireless sensor networks, cloud computing, machine to machine, as shown in Figure 1.7.

Green Internet Technologies

Green Internet Technologies require special hardware and software which are specifically designed to consume less energy without reducing performance, thus minimizing utilization of resources.

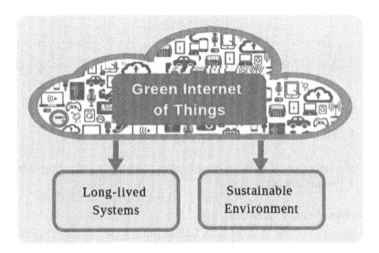

FIGURE 1.6 Green IoT environment

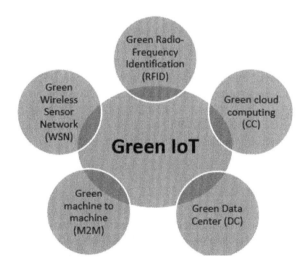

FIGURE 1.7 Green IoT communication components

Green RFID Tags

RFID tags can store data or information at a small level for any objects with which these are linked. RFID transmission requires a few meters' range of RFID systems. One type of RFID tag, Active tags, has built-in batteries for continuously transmitting their own signal while another type, Passive tags, doesn't have an active battery source. These store energy from the reader. Reducing the size of an RFID tag can help in reducing the amount of non-degradable material because it is not an easy task to recycle tags. Energy-saving techniques and protocols must be utilized for tag

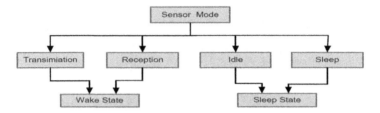

FIGURE 1.8   Sensor modes for Green IoT

collision avoidance, tag estimation, overheating avoidance, adjusting transmission power level dynamically, etc.

Green Wireless Sensor Network

WSN has many sensor nodes but with limited power and storage capacity. Green WSN can be achieved by green energy conservation techniques, radio optimization techniques, green routing techniques, leading to reduction in energy consumption. Smart data algorithms will reduce storage capacity requirements, as well as data size, and activate sensors as and when required. The principle of green IoT is enabled by having the sensor nodes in sleep mode to save energy for most of their lives. Only when data communication takes place at ultra-low power, can a WSN be realized. Sensors may make use of directly extracted energy from the atmosphere, such as light, vibrations, kinetic energy, temperature differentials, etc. WSN technology must efficiently transmit a signal and allow sleeping for minimal power usage as shown in Figure 1.8. In sensors microprocessors must also be able to smartly wake up and sleep. Microprocessor patterns for WSNs therefore include reducing energy consumption, thus increasing the speed of the processor.

Green Cloud Computing

Cloud computing has services like IaaS, PaaS, and SaaS. In a green cloud computing scheme, hardware and software are to be used in such a way as to reduce energy consumption. Policies have to be applied which are energy efficient. Also, green cloud computing scheme based technologies are to be used like communication, networking, etc. The primary aim of green cloud computing is to encourage the use of biodegradable products which can be easily recycled and reused, as shown in Figure 1.9.

Green Data Centers

Data centers are responsible for storing, managing, processing, and disseminating all types of data as well as applications. Data centers should be designed which are based on renewable energy resources. Apart from this, routing protocols should be designed to make them energy aware, turn off idle devices in the network, and incorporate energy parameters for their routing of packets.

Green M2M

Since a large number of machines are involved in M2M communication, there should be energy saving based transmission power and optimized communication protocols,

FIGURE 1.9 Green cloud computing

and routing algorithms. There should be check passive nodes so that energy can be saved.

1.3 APPROACHES FOR GREEN IOT

There is a need to research more state-of-the-art technologies and methods to make the IoT green and satisfy the energy hunger of billions of users. Green IoT is attracting the researchers working in IoT area as conventional energy supplies are speedily declining and the requirement for energy is growing exponentially. This section broadly classifies major approaches for green IoT as shown in Figure 1.10.

1.3.1 HARDWARE BASED GREEN IoT TECHNIQUES

These techniques require designing hardware (integrated circuit (IC), processor, sensor, RFID, etc.) which is energy efficient and has minimum negative impact on the environment. IC architecture within an IoT network is critical for energy preservation.

A Green Sensors on Chip (SoC) model enhances the architecture of IoT networks by integrating sensors, processing power on a single chip to reduce traffic, e-waste, carbon footprint and overall infrastructure energy consumption. While, the example of Sleep Walker illustrates energy conservation using Green SoC, more energy can be conserved for this model using recyclable material. Likewise, CoreLH, an IoT dual core energy efficient processor,

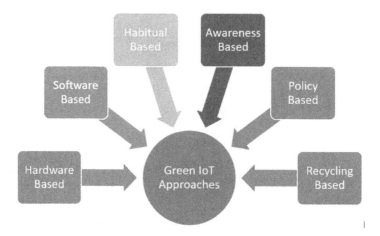

FIGURE 1.10 Approaches for Green IoT

has a CoreL for low-computation tasks and CoreH for high-computing tasks. This lowers energy consumption through a scheduling framework that assigns different tasks to these cores on the basis of the resources they may require.

1.3.2 SOFTWARE BASED GREEN IOT TECHNIQUES

These techniques require writing software for data centers, cloud computing, and virtualization which can assist IoT techniques to perform better in terms of energy efficiency. Sensors use energy while they are idle but turned on, thereby wasting a lot of energy. Therefore, an energy-efficient scheduling algorithm that adjusts the sensor status to on-duty, pre-off-duty, and off-duty depending on the situation is required to avoid unnecessary use of energy.

A Mixed Integer Linear Programming (MILP) virtualization system proposed to provide a four-layer architecture in which IoT devices are put in upper layers and networking components in lower layers. Results show that 36% less energy is consumed by applying this framework. A Programming Language named EPDL was designed to help the non-experts to write energy policies for a smart environment like IoT. Energy is wasted in the RFID as the tags sit idle but are turned on, as they don't know when to communicate beforehand. This is called the overhearing problem which is responsible for an enormous waste of energy resources. To solve this, a method was proposed for measuring the times of contact in advance and for putting the tags to sleep when not in use. Although this model has saved a lot of energy, but additional resources are lost in deciding contact time and shifting states from ON to Sleep and vice versa.

1.3.3 HABITUAL BASED GREEN IOT TECHNIQUES

Another approach that can be implemented for achieving energy efficiency and reducing carbon emissions is to follow simple practices through which we can reduce energy use in our daily activities. Although the measurement scale is very small, still it can make a huge difference if we add up the small savings on a world-wide scale.

One approach is to monitor energy consumption patterns by automation systems in workplaces, residences, industries, and then reduce energy losses in our routine activities.

1.3.4 AWARENESS BASED GREEN IoT TECHNIQUES

Spreading awareness is one of the vital factors in reducing energy consumption. This is highly dependent on factors like region, culture, place, education level, as it is not precisely predictable whether people will pay attention to awareness movements and follow official advice. Therefore, "smart metering technology" may be used to deliver to homeowners real-time data on their energy ingestion from different sources in their residences, workplaces, buildings, and then we can advise them on how to monitor and mitigate their energy consumption based on that real-time information. This can save 3–6 percent of energy.

1.3.5 POLICY BASED GREEN IoT TECHNIQUES

Policies and policies focused on IoT applications' real-time data will aid in large-scale energy savings. There are different stages in the implementation of energy efficiency policies such as control (different energy use situations), information management, user feedback, and automation programs. Data can be collected from various parts of a building where the behavior of the residents varies, and energy ingestion differs. These data can be analyzed to formulate policies and strategies for different parts of the same building depending on their energy needs.

1.3.6 RECYCLING BASED GREEN IoT TECHNIQUES

Use of recyclable material in an IoT network for the development of products will help make it an environment-friendly one. For example, cell phones are made of some of the scarcest natural resources, such as copper, and consist of some non-biodegradable elements, such as plastic, that can increase the greenhouse effect if not handled properly when the phones are no longer in use.

An estimate indicates that 23 million mobile phones are present in Australia's drawers and cupboards that are no longer in use and that 90 percent of the content in the phones is recyclable. Recycling is needed if we're going to tackle the greenhouse effect issue and huge energy use. While recovering 90 percent of material is an unreasonable expectation, recycling still can make a difference. Some of the green IoT techniques for waste management are described in Figure 1.11.

Green IoT requires wireless sensors to help track fill level in waste containers by waste companies and to automate waste collection routes. As of now, the Enevo sensors are reducing the cost of waste collection by about 50 percent – the company now operates in 36 countries. The smart containers compress the waste automatically, raising the collection frequency by around 70 percent on average.

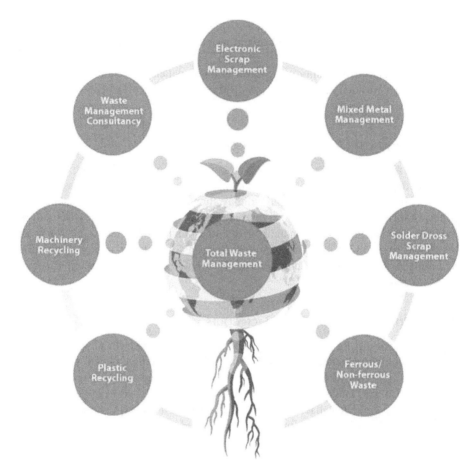

FIGURE 1.11   Recycling based Green IoT techniques

1.4   PRINCIPLES FOR GREEN IOT

Based on the conceptual and empirical literature, there are five basic principles to accomplish Green IoT and mitigate carbon emission (depicted in Figure 1.12).

- Reduced Network Size: Increase network size by using ingenious routing methods and effective node positioning. This will contribute to energy savings on the high end.
- Using Selective Sensing: Collect only the data required in a given situation. Eliminating extra sensing of the data can save a lot of energy.
- Use of Hybrid Architecture: The use of Passive and Active sensors for various types of tasks in an IoT network will reduce energy use.
- Policymaking: Build efficient energy-reduction strategies in smart buildings. Policies can have a direct impact on energy consumption and can thus save a significant amount of energy.

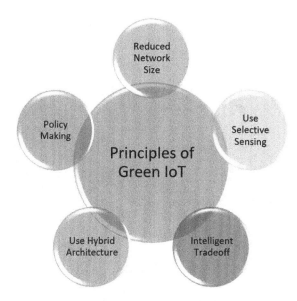

FIGURE 1.12   Principles of Green IoT

- Intelligent Trade-Offs: We have to trade off everywhere so we can smartly allo-
  cate costs and save resources in certain situations, such as compressive sensing
  and data fusion, in processing or communication. Trade-offs will be selected
  based on a specific scenario.

Some additional principles can be summarized as:

- Turn off services and facilities that are not needed (e.g. "sleep scheduling")
- Only send required data (e.g. "predictive data delivery")
- Focus on minimizing data path length (e.g. "routing schemes")
- Adopt trade-off processing for communication (e.g. "data fusion")
- Utilize advanced communication technologies (e.g. "MIMO")
- Make use of renewable green power sources (e.g. "solar energy").

Eco-friendly designs make use of both natural and green light, which means
electricity generated by solar or wind power to operate the data center. Embracing
eco-friendly designs offers companies benefits such as complete design power, heat
reduction, and lighting.

1.5   APPLICATIONS OF GREEN IOT

Applications of Green IoT and IoT are not different, in fact they are the same. Some
of the main applications are Smart City, Smart homes, Smart market, Smart Grid
Systems, Smart Infrastructure, Smart Factories, Smart Medical Systems and Smart
Logistics similar to IoT. Applications of Green IoT focus more on energy conservation

FIGURE 1.13   Application of Green IoT for smart world

aspects and the goal of reducing emissions from sensor and IoT devices, leading to a smart world as depicted in Figure 1.13 and 1.14.

In the current scenario, our cities are facing grim problems such as a high carbon footprint, inefficient transportation systems, ineffective healthcare, and other public services. Most of these predicaments can be related to the higher rate of migration to the cities. A smart city can be an effective answer to these problems. The concept of smart city is explained by many researchers, interest groups, or stakeholders on the basis of their interpretation of the term "smart," and therefore there is no single globally accepted definition. The general impression is that a smart city is a city where technology and innovation are used for betterment of the quality of life by optimizing resources and by improving infrastructure and public services.

Washburn et al. defined a smart city as "The use of smart computing technologies to make the critical infrastructure components and services of a city – which include city administration, education, healthcare, public safety, real estate, transportation, and utilities – more intelligent, interconnected, and efficient" [3].

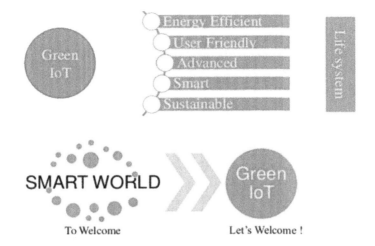

FIGURE 1.14   Green IoT leads to smart world

Further:

> We believe a city to be smart when investments in human and social capital and traditional (transport) and modern (ICT) communication infrastructure fuel sustainable economic growth and a high quality of life, with a wise management of natural resources, through participatory governance.

Geller defined a smart city as "A city well performing in a forward-looking way in economy, people, governance, mobility, environment, and living, built on the smart combination of endowments and activities of self-decisive, independent and aware citizens" [6]. IDC Government Insights provides a smart city maturity model as depicted in Figure 1.15.

It describes the stages of development of a smart city. A city cannot be simply smart or dumb. This model states that a city can be in any of the stages, i.e. ad-hoc, opportunistic, repeatable, managed, optimized.

i.   *Ad-hoc:* at this level government does not have any formal process to implement new concepts.
ii. *Opportunistic:* at this level experimentations are done with people by means of mobile apps and community networking.
iii. *Repeatable:* this level encourages recurring communications between citizens and governments.
iv. *Managed:* at this level innovation is organized.
v.   *Optimized:* this is the highest maturity level a city can achieve; this refers to the state where the government has perfected innovation models.

A city can start from ad-hoc level and can gradually attain the status of optimized. A smart city must have these six characteristics (as shown in Figure 1.16).

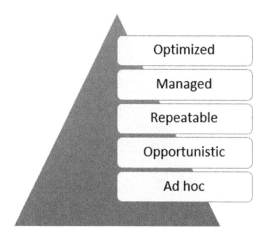

FIGURE 1.15   Smart city maturity model

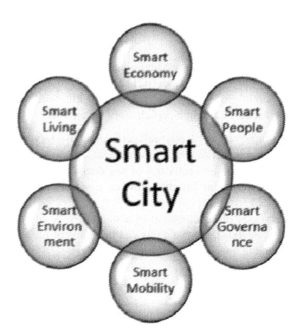

FIGURE 1.16   Characteristics of smart city

- *Smart Economy:* This includes the factors that result in all-around economic growth of the city. This specifically refers to improving innovation and productivity and growth in field of Information and Communication Technology.
- *Smart People:* Smart people refers to the fact that citizens in a smart city not only should be achieving higher academic qualifications but they should be highly capable in interaction within the community and with the outer world.

- *Smart Governance:* This generally refers to e-governance, which aims at better association between administration and the citizens, also at delivering public services to its citizens efficiently.
- *Smart Mobility:* It explains the need and solution for smart ways to deploy transportation resources, better traffic management, and effective parking solutions.
- *Smart Environment:* This points the factors for reduction of pollution and providing better natural living conditions.
- *Smart Living:* It describes the factors for achieving higher quality of life, such as safety, education, and cultural facilities, social interaction events, healthcare services, etc.

According to the Ministry of Urban Development, Government of India, the core infrastructure in a smart city should include the elements displayed in Figure 1.17.

These guidelines also provide smart solutions to achieve the vision of converting a city into smart city. For good governance, e-governance should be opted, so that all the public domain services are easily accessible to the citizens and transparency can be assured. For efficient waste management, the waste should be properly collected

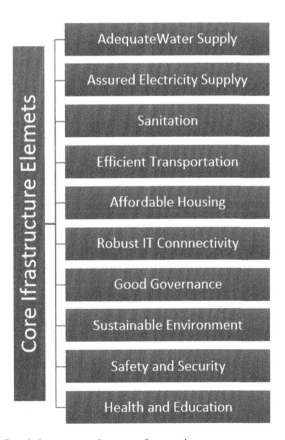

FIGURE 1.17 Core infrastructure elements of smart city

and recycled, organic waste should be converted to energy, fuel, or compost. Waste water should be treated effectively. Water supply management should be through smart meters with water quality monitoring and leakage identification and prevention. For efficient energy management, renewable sources of energy should be adopted; energy efficient architecture should be used for building construction. For efficient transportation, intelligent transportation management, smart parking, and integrated multi-modal transport should be implemented.

A smart city must aim to monitor its resources like various transportation modes, fluid supply system, buildings, and its communications system using advanced sensors. Only then can its resources be efficiently deployed and the city be cost effectively managed. The concept of smart cities cannot flourish without the help of industry giants. Some of the tech industry titans have been working to realize smart cities in the real world. CISCO and IBM have been contributing to making the concept of smart city reality. IBM states that infrastructure and operations along with people provide the base for the concept of smart city. All the services that make a city "smart" can be classified into three categories services concerning individuals, infrastructure related services, and last but not the least planning and administrative services.

The characteristics of a smarter world can be stated as:

i.   Instrumented: We can sense, measure, and record everything
ii. Interconnected: Everything is connected and can communicate with each other (the concept of IoT)
iii. Intelligent: By analyzing quickly and accurately the data gathered, we can predict and prepare ourselves for forthcoming events.

The smart services of a smart city are

i.   Smart energy metering: for efficient delivery and lowering energy consumption by analyzing the usage patterns
ii. Smart fluid delivery system: for monitoring of water usage and quality, efficient water distribution
iii. Air quality monitoring system: for real time monitoring of air pollution levels, harmful gas detection, humidity and temperature change detection
iv. Condition monitoring of buildings and structures: for monitoring, detection, and locating damage to them.

The limiting factors during implementation of smart cities are:

1.   Increasing Urban Population: The population growing at such an alarming rate and high migration rate to the cities poses a critical challenge to the realization of smart cities, as with the increase in population the implementation and integration of various services becomes more and more difficult.
2.   Higher Cost of Implementation: At the current stage, the components required and the services necessary for smart cities are expensive. This results in higher costs of implementation.

3. Less R&D: This is a relatively new concept and there has been comparatively less research in this field. Exhaustive research in any the field will also result in efficient implementation.

4. Security and Privacy Concerns: As the implementation of the smart city relies on networks, it has all the flaws and limitations of networks. For better results, new networking technologies and better network protocols must be designed.

5. Traffic Congestion: Traffic congestion poses a grave concern for an efficient transportation system. The transport system is highly overloaded, and a significant amount of money is wasted due to congestion every year.

6. High Energy Consumption: Energy consumption in the average home or office is relatively high and can be reduced if smart solutions are implemented.

7. Dependence on Conventional Energy Sources: We have been relying on conventional/non-renewable sources for a very long time. For smart environment, we have to switch to other options very soon.

8. Healthcare Limitations: Healthcare is one of the most important and expensive sectors. Healthcare services are either not accessible to a large portion of population or they are pricey in most countries. E-health can be a better solution in these conditions.

9. Large Amount of Data: While implementing smart cities, we have to deal with a large amount of data, and so new techniques have to be devised for its analysis and processing.

Smart cities propose a better living environment with a higher living standard. A city does not become "smart" only by technological advancements, but it is collective growth of infrastructure, the community, and the citizens residing in the city. Any city can be transformed into a smart city by following the guidelines proposed by smart city maturity model. Furthermore, a smart city should opt for smart solutions for healthcare, transport, and other public services. In smart cities, objects can interact smartly with people via the IoT. By detecting emissions through IoT and environmental sensors, smart cities can become a greener area. Green IoT ensures the sustainability of green places in smart cities. Green IoT makes smart cities safe automatically and smartly. Governments and many organizations around the world are making a lot of efforts to promote the reduction of energy usage, carbon emissions, as well as promoting the Green IoT for smart cities.

1.6 FUTURE RESEARCH DIRECTIONS OF GREEN IOT

The enormous growth of IoT technology has changed the way we live and work. While the various benefits of IoT enrich our environment, it should be noted that the IoT still is energy intensive, produces harmful gases, and e-waste. This puts new pressures on the smart world and the climate. Future research about green IoT must address the issues described in Figure 1.18.

There is a growing desire to move towards green IoT in order to boost profits and diminish the damage caused by IoT. Green IoT is seen as the environmentally friendly future of IoT. To achieve this, some essential steps are required to be followed to reduce carbon emissions, waste fewer resources, and encourage usage of renewable

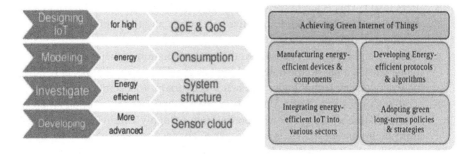

FIGURE 1.18   Research focus of Green IoT

energy resources. This is the reason for moving towards green IoT, where every IoT device or node is energy efficient with minimal carbon emissions.

BIBLIOGRAPHY

[1]   R. Arshad, S. Zahoor, M. Ali Shah, A. Wahid, H. Yu, Green IoT: An investigation on energy saving practices for 2020 and beyond. *IEEE Access,* 5 (2017): 15667–15681.

[2]   M. Maksimovic, Greening the future: Green Internet of Things (G-IoT) as a key techno-logical enabler of sustainable development. In Dey N., Hassanien, A., Bhatt, C., Ashour, A., and Satapathy, S. (eds) *Internet of Things and Big Data Analytics toward Next-Generation Intelligence.* Studies in Big Data, vol. 30. Springer, Cham, 2018.

[3]   D. Washburn, U. Sindhu, S. Balaouras, R.A. Dines, N. Hayes, and L.E. Nelson. Helping CIOs understand "smart city" initiatives. *Growth* 17 (2009): 1–17.

[4]   A. S. H. Abdul-Qawy and T. Srinivasulu, Greening trends in energy-efficiency of IoT-based heterogeneous wireless nodes. IEEE International Conference on Electrical, Electronics, Computers, Communication, Mechanical and Computing (EECCMC-2018)., Vellore, Tamil Nadu, 2018.

[5]   A. Ozturk, I. Togla, I. Tolga Medeni, T. Burak Ucuncu, M. Caylan, F. Akba, T. Durmus Medeni, Green ICT (information and communication technologies): A review of academic and practitioner perspectives. *International Journal of eBusiness and eGovernment Studies,* 3(1) (2011): 1–16.

[6]   A. L. Geller, Smart growth: A prescription for livable cities. *American Journal of Public Health*, 93 (2003): 1410–1415.

[7]   A. Al-Fuqaha M. Guizani, M. Mohammadi, M. Aledhari, M. Ayyash, Internet of things: A survey on enabling technologies, protocols, and applications. *IEEE Communications Surveys and Tutorials,* 17(4) (2015): 2347–2376.

[8]   S. Alsamhi, M. Samar Ansari, Q. Meng, Greening internet of things for smart everythings with a green-environment life: A survey and future prospects. May 2018. *arXiv preprint arXiv:1805.00844.*

[9]   R. Ahmed, Khan, P. Farooq. Green IoT: Issues and challenges. 2nd International Conference on Advanced Computing and Software Engineering (ICACSE-2019).

[10] L. Atzori, A. Iera, and G. Morabito, The internet of things: A survey. *Computer Networks*, 54 (2010): 2787–2805.

[11] J. Gubbi, R. Buyya, S. Marusic, and M. Palaniswami, Internet of Things (IoT): A vision, architectural elements, and future directions. *Future Generation Computer Systems,* 29 (2013): 1645–1660.

[12]   S. S. Prasad and C. Kumar, A green and reliable internet of things. *Communications and Network*, 5 (2013): 44.

[13]   C. Zhu, V. C. Leung, L. Shu, and E. C.-H. Ngai, Green Internet of Things for the smart world. *IEEE Access*, 3 (2015): 2151–2162.

[14]   S. Sala, Information and communication technologies for climate change adaptation, with a focus on the agricultural sector. Thinkpiece for CGIAR Science Forum Workshop on "ICTs transforming agricultural science, research, and technology generation," Wageningen, Netherlands, 2009, pp. 16–17.

[15]   H. Eakin, P .M. Wightman, D. Hsu, V .R. Gil Ramón, E. Fuentes-Contreras, M. P. Cox, T.-A. N. Hyman, C. Pacas, F. Borraz, and C. González-Brambila, Information and communication technologies and climate change adaptation in Latin America and the Caribbean: a framework for action. *Climate and Development*, 7 (2015): 208–222.

[16]   A. P. Upadhyay and A. Bijalwan, Climate change adaptation: Services and role of information communication technology (ICT) in India. *American Journal of Environmental Protection*, 4 (2015): 70–74.

[17]   N. Zanamwe and A. Okunoye, Role of information and communication technologies (ICTs) in mitigating, adapting to and monitoring climate change in developing countries. International conference on ICT for Africa, 2013.

[18]   I. Mickoleit, *Greener and Smarter: ICTs, the Environment and Climate Change* OECD Publishing, Paris, 2010.

[19]   A. L. Di Salvo, F. Agostinho, C. M. Almeida, and B. F. Giannetti, Can cloud computing be labeled as "green"? Insights under an environmental accounting perspective. *Renewable and Sustainable Energy Reviews*, 69 (2017): 514–526.

[20]   P. Morreale, F. Qi, and P. Croft, A green wireless sensor network for environmental monitoring and risk identification. *International Journal of Sensor Networks*, 10 (2011): 73–82.

[21]   A. Liu, Q. Zhang, Z. Li, Y.-j. Choi, J. Li, and N. Komuro, A green and reliable communication modeling for industrial internet of things. *Computers and Electrical Engineering*, 58 (2017): 364–381.

[22]   J. Shuja, R. W. Ahmad, A. Gani, A. I. A. Ahmed, A. Siddiqa, K. Nisar, S. U. Khan, and A. Y. Zomaya, Greening emerging IT technologies: Techniques and practices. *Journal of Internet Services and Applications*, 8 (2017): 1–11.

[23]   J. Li, Y. Liu, Z. Zhang, J. Ren, and N. Zhao, Towards Green IoT networking: Performance optimization of network coding based communication and reliable storage. *IEEE Access*, ? (2017): 8780–8791.

[24]   www.techiexpert.com/green-iot-way-to-save-the-environment/

[25]   www.slideshare.net/TAPASKUMARPAUL2/presentation-on-green-iot

[26]   C. Zhu, C. Victor, M. Leung, E. Ngai, Green Internet of Things for smart world. *IEEE Access*, 3 (2015): 2151–2162.

[27]   www.sketchbubble.com/

2 Green Internet of Things (G-IoT)

A Solution for Sustainable Technological Development

Puneet Garg, Shrivastava Pranav, and Agarwal Prerna

CONTENTS

2.1 INTRODUCTION

The use of computationally advanced devices, smartphones, etc. is increasing at an unprecedented rate. Large numbers of users and their multiple devices have led to alarming levels of power utilization over the last decade. By a calculated assumption, connected devices by 2020 will be around 50 billion [1], and 100 billion by 2030 [2]. Researchers anticipate that this will require a massive data rate and a large content size (tens of thousands of times higher than compared with 2010) at the expense of unprecedented amounts of carbon emissions. It is expected that by 2020 cellular networks will lead to more than 345 million tons of CO_2 production, and it is estimated that it will intensify in the forthcoming years [3]. Full pollution forecasts are given for 2020 [4]. Because of these enormous pollution, environmental and medical issues, green or sustainable technology has emerged as an important area of investigation in technological progression. Additionally, current device battery technology is an additional chief apprehension that spearheads the green technology [5].

Researchers predicted that from mid-2020, the fifth generation of wireless communications (5G), capable of handling nearly thousand times more mobile data than current cellular systems will be fully deployed [6]. There are five powerful 5G communication innovations, as depicted in Figure 2.1. IoT's functionality permits the binding of billions of users within a small-time interval. By reducing latency, Device-to-Device (D2D) connectivity increases the speed and reliability of user-to-user communication and Spectrum Sharing (SS) is aimed at preventing the impact of poor spectrum usage. Furthermore, the UDNs involve massive small-cell distribution, in places of large traffic. In comparison, the huge MIMO offers a high data rate by allowing hundreds of antennas. These influential innovations will enable the energy consumption of the upcoming 5 G networks to be regulated to reduce the (CO_2) emissions. The main objective of this chapter is providing summary of definition, implementations, innovations and challenges of green IoT.

2.2 INTERNET OF THINGS

In 1999, one affiliate of the Radio Frequency Identification development community (RFID DevCom) conceptualized the Internet of Things (Figure 2.2). The escalation in mobile devices usage supported by omnipresent communication, increasing computation power and data analytics have made it more relevant to the practical world. Envisage an integrated scenario where billions of devices will be capable of sensing, communicating and sharing information. Such integrated structures include data collection, processing and usage to facilitate regular actions, supplying strategy, control and administration with a range of information. This is Internet of Things (IoT) culture [7]. The public notion of IoT is:

> Internet of Things (IOT) is a network of physical objects. The Web is not only a network of computers, but it has grown into a network of devices of all shapes and sizes, cars, tablets, home appliances, toys, cameras, medical devices and industrial systems, livestock, humans, houses, all linked, all interacting and sharing information on the basis of protocols to accomplish smart reorganizations, mapping, tracking, surveillance. [1] [2]

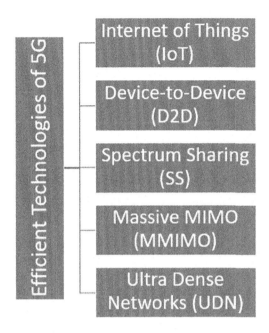

FIGURE 2.1 Efficient Technologies of 5G

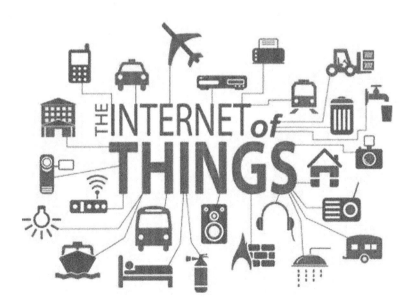

FIGURE 2.2 Internet of Things [10]

The Internet of Things is a three-dimensional internet, we describe IoT as follows in three categories:

1. Users to users
2. Users to devices
3. Devices to devices.

IoT is a notion and a standard that contemplates comprehensive existence in the world of innumerous devices that are capable of interacting among themselves, through wired and wireless links, and special addressing schemes, and collaborate with other devices to create new applications/services and accomplish shared objectives. The research and development complexities in developing a sustainable environment in this sense are massive, with a future in which physical, digital, and virtual converge together to build smart worlds [8] [9].

IoT relates to the common outline of things, in particular general objects, which are visible, identifiable, traceable, and addressable by data-sensing devices and/or manageable via the internet, regardless of connectivity mediums (whether via RFID, WLAN, WANs, or further channels). Everyday objects comprise electronic devices we use in our routine procedures, advanced technical progress like equipment and vehicles, and, in addition, objects we don't usually perceive to be electronic, like clothes, food, animals, etc. [8] [9].

IoT is a new cyber movement where devices are identifiable and gain wisdom as a result of taking or permitting circumstantial judgments, as they are capable of conveying knowledge regarding them. Any device is able to access data aggregated by other entities, or they may become elements of structured systems. This revolution corresponds with cloud computing advancements and the internet's move to IPv6 with limitless addressing power [8] [9]. The aim of the Internet of Things is to provide the ability to link objects anywhere, wherever, with anyone and everyone preferably exploiting whichever network or whichever device, as depicted in Figure 2.3.

2.2.1 The IoT Technology Stack

If anyone wants to navigate the IoT engineering labyrinth, due to the complexity and sheer numbers of technology solutions that accompany it, it can be a tough task. However, we can break down the stack of IoT technologies into four simple levels of technology concerned with driving IoT.

a. Device hardware: Devices are artifacts that are essentially the "things" within the IoT. They act like an intermediary between the digital and physical infrastructures, assuming diverse dimensions and complexities based upon their role as per requirement, contained by the particular IoT implementation. Miniscule spy cams or large structural devices, nearly any object (humans or animals) may be converted as a connected device via installing required instrumentation (through attaching actuators or sensors together accompanied by the correct software) to calculate and accumulate the needed data. Actuators, sensors, or else supplementary telemetry devices might,

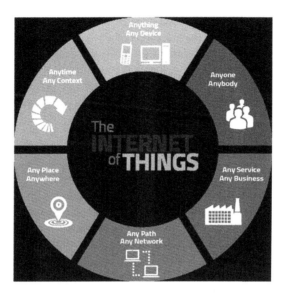

FIGURE 2.3 Internet of Things

of course, also perform as stand-alone smart devices. Individual drawback found is genuine case for IoT usage and hardware specifications (dimensions, simplicity of installation and maintenance, durability, helpful service life, cost-effectiveness).

b. Device software: Software is accountable for realizing internet connectivity, data collection, system deployment, and data monitoring within the IoT network. Additionally, it also provides consumers with the computer functionality to imagine data and communicate with the IoT framework with application-level capabilities.

c. Communications: Although it is possible for communications frameworks to be strongly coupled with computer software and hardware, treating them as a separate layer is important. Communication framework contains all physical communication mechanisms (cellular, cable, LAN) and common protocols used in different IoT environments (ZigBee, Z-Wave, Thread, MQTT, LwM2 M). Selecting the right connectivity approach is an essential component in constructing a stack of IoT technologies. The selected technologies would decide not only cloud communication protocols, but systems handling and interaction with third parties' apps.

Platform: As suggested, a computer can "feel" what's going on around it thanks to the smart hardware and the enabled apps, and relay that to the consumer through a different communication channel. The IoT framework is the location to capture, organize, store, interpret and display all of these details in a comprehensible and intelligible manner. This renders such a system particularly beneficial; however, it is not only its data collection and processing capability, but its capacity for analyzing and finding useful information from the data portions generated as a result of the devices interaction by communication layer. However, there is rather a range of IoT solutions available in the marketplace, having preference based on the particular IoT project specifications and considerations such as stack design and IoT technologies, usability, configuration properties, common protocols, device agnosticism, protection and cost-effectiveness. In addition, on-premises or cloud-based systems can also be built, e.g., Coiote IoT Device Management is capable of deployment both on-site and within cloud.

2.2.2 Connectivity Options Inside the IoT Stack

The problem related to shortage of connectivity solutions does not hinder real-life applications of the IoT technologies. The question is, can a communications solution offer different service enabling scenarios based upon requirements of provided IoT use-case while providing balance between energy demand, coverage and bandwidth. For example, suppose someone is creating a smart home; he might wish for internal temperature sensors and a heating controller assimilated with smartphones so he may supervise the temperature of every room remotely and adjust them to the current needs. In such cases, the recommended solution would be the Thread, an IPv6 networking protocol, specifically premeditated for home-automation system.

This complexity and diversity related to networking norms and protocols, raises another issue regarding actual requirement to develop innovative implementations even as a number of Internet protocols already exists from decades. This is happening because accessible communication protocols, such as Transmission Control Protocol/Internet Protocol (TCP/IP), are not adequately powerful and require more energy for working proficiently in evolving IoT implementations. Each segment will provide a brief overview of the main alternate internet protocols dedicated specifically to IoT devices.

The summary discusses the most common IoT network applications, broken down by the spectrum of radio frequencies reached through these solutions: short-range solutions, medium-range solutions, and long-range solutions.

Short-Range IoT Network Solutions

a. **Bluetooth** It is a proven short-range communication protocol; and is regarded as a main remedy especially in support of prospects of wearable electronics market like wireless headphones or geolocation sensors, particularly due to its universal inclusion in smartphones. The Bluetooth Low-Energy (BLE) protocol is deliberated with practical and reduced energy consumption and necessitates incredibly minute power from a device. However, there exists a downside: this protocol isn't the best practical solution while communicating bigger data quantities regularly.

b. **Radio Frequency Identification** As Radio Frequency Identification (RFID) is a pioneering IoT technology, it offers promising IoT solutions, particularly for logistics and SCM that involve the capability to determine entity location within buildings. Clearly, RFID prospects surpass uncomplicated localization services, having possible implementations varying from managing patients to augmenting the efficiency of healthcare services to furnishing real-time inventory data to mitigate out-of-stock situations for outlets.

Medium-Range Solutions:

a. **Wi-Fi** It is established on IEEE 802.11, and continues to be the most extensively used and commonly recognized protocol for wireless communication. The extensive use throughout the IoT environment is primarily restricted due to elevated energy consumption because of the requirement to maintain signal strength and swift data transmission in favor of enhanced efficiency and connectivity. IT offers an extensive range for surprising numbers of IoT applications; nevertheless, it also necessitates management and regulation in marketing to deliver revenues to both consumers and service providers. Linkify is one good example of a Wi-Fi management platform that provides a value-added service that empowers public Wi-Fi access points. As one of AVSystem's cutting-edge technologies, Linkify provides Wi-Fi configuration and communication opportunities for practically limitless visitors.

b. **ZigBee** It sees its largest use in household electronics, traffic management systems and the machinery industry. It is developed on the standard IEEE 802.15.4. Zigbee facilitates low data exchange rates, low power service, safety and reliability.

c. **Thread** It is premeditated exclusively for smart home items and uses IPv6 networking to allow connected devices for interacting among themselves, access cloud services, or provide consumer interaction through Mobile Thread applications. Thread's analysts have suggested that with the competitiveness of the industry, yet another protocol for communication is contributing to supplementary disintegration inside IoT stack.

Long-Range Wide Area Networks (WAN) Solutions

a. **NB-IoT** It is the latest standard of radio technology that guarantees exceptionally low energy utilization (battery power of ten of years) and supplies connectivity with an approximate signal strength, 23 dB smaller than in 2 G. It also utilizes accessible communication framework that imparts global coverage of LTE networks and in addition boosts signal efficiency. The mentioned advantage allows NB-IoT to be implemented instead of solutions that require local network building, such as Sigfox, LoRa etc.

b. **LTE-Cat M1** LTE Cat M1 is a standard for low-power wide-area (LPWA) connectivity that connects IoT and M2 M devices with medium data requirements. This facilitates longer battery life cycles and has an increased range of building relative to wireless systems such as 2 G, 3 G, or LTE Cat 1. Because CATM1 is compliant with the current LTE network, it does not allow the carriers to build new networks for deployment. Compared with NB-IoT, LTE Cat M1 appears to be ideal for mobile usage cases, since its handling of managing between cell sites is significantly better and very close to high-speed LTE.

   c. **LoRaWAN** It is a low power Long Range Wide-Area Networking protocol optimized for low power consumption and with millions of devices supporting large networks. LoRaWAN is designed to provide low-power WANs with features required to support low-cost, mobile, and secure bi-directional communication in IoT, M2 M, smart city and industrial applications for wide-area (WAN) applications.

   d. **Sigfox** The idea behind Sigfox is to provide an efficient networking approach for low-power M2 M applications that require low data transmission rates for which the Wi-Fi range is too limited, and the cellular range is too costly and too power-hungry. Sigfox uses UNB, a platform that helps it to accommodate low data transfer speeds ranging between 10 and 1,000 bits per second. Consuming up to 100 times less energy than wireless connectivity systems, this offers a standard 20-year stand-by cycle for a 2.5Ah pack. Offering a reliable, energy-efficient and flexible network capable of supporting connectivity between thousands of thousands of battery-operated devices over several square kilometers of land, Sigfox is ideal for numerous M2 M applications including smart street lighting, smart meters, patient alarms, safety devices and environmental sensors. Sigfox is currently employed in an increasing number of IoT technology solutions, such as the Coiote IoT Software Orchestration by AVSystem, to name just one.

2.2.3   ENABLING TECHNOLOGY FOR IoT

IoT is a universal information infrastructure that permits highly developed services through interconnecting (virtual and physical) objects/devices centered upon accessible and developing ICT. With IoT the network is applied to all the objects/devices surrounding us via the Web. IoT is so much more than just machine-to-machine connectivity, cellular sensor networks, sensor networks, GSM, GPRS, RFID, WI-FI, GPS, microcontroller, microprocessor, etc. These are the underlying technologies that make "Internet of Things" feasible. The enabling technologies for Internet of Things and can be divided into three categories [8]:

1. Technologies enabling "things" to collect relative and appropriate data,
2. Technologies enabling "things" to handle contextual information; and
3. Technologies to enhance privacy and security.

The first two definitions can be interpreted together as conceptual building blocks that allow the integration of "intelligence" into "things," which are indeed the characteristics that separate the IoT from the ordinary internet. The last category is not a requirement related to functionality; instead, it is an imperative and genuine prerequisite devoid of which the IoT penetration might be unrelentingly diminished [2]–[9]. IoT is often promoted as a standalone technology. On the contrary, it is a combination of several software and hardware innovations. IoT offers resolutions centered on IT integration that purport to software and hardware exploited for storing, retrieving, and processing information, and networking, which comprises electronic systems intended for group or individual interaction.

The diverse combinations offered by current networking technologies necessitate modification for becoming appropriate for requirements of IoT applications like efficient power consumption, reliability, security and speed. Pertaining to this, it's likely that diversification may remain limited, towards a range of achievable networking solutions that deal with requirements of IoT implementations, are embraced by the industry; already proved to be serviceable, and backed by a broad development partnership. Definitions of such specifications cover wired and wireless systems such as GPRS, GSM, Bluetooth, WI-FI, and ZigBee [8] [9].

2.2.4 IoT Characteristics

The following characteristics are IoT basic features [9] [12]:

a. **Interconnectivity:** any device capable of integrating with a global communication network will be able to so for implementing IoT.

b. **Things-related services**: IoT is competent to deliver services related to objects/devices, contained by restraint of things, like confidentiality and functional continuity of material things and their abstract corresponding items. For providing object-related services within the constraint of things, both the knowledge environment and physical world technology will be evolving.

c. **Heterogeneity:** Amalgamation from various networks and platforms, leads to heterogeneity in IoT systems. These systems might correspond, with other devices or application platforms, within several networks.

d. **Dynamic changes:** System state varies continuously, e.g., powering-up and down, linking or/and disconnecting, and system dimensions like position and velocity. In fact, the device's number of might alter vigorously.

e. **Enormous scale:** The quantity of devices that needs supervising and that contact all others will be greater than the devices connected to the current internet. The control of the produced data and their analysis for the purposes of implementation will be even more important. It includes data semantics, as well as secure data handling.

f. **Safety:** As IoT provides benefits, safety as an imperative issue must not be ignored. IoT developers and users both have to plan for protection. It requires the protection of our personal data and our physical wellbeing. Safeguarding data and networks necessitates developing a scalable security model.

g. **Connectivity:** Connectivity enables compatibility and accessibility of networks. Compatibility provides the ability to generate and consume information, while accessibility is ability to access a network.

2.2.5 Architecture of IoT

There is no overarching standard for IoT design that is universally agreed. Several scholars have suggested various architectures. The primitive architecture is an architecture with three layers [13] [14] [15] as depicted in Figure 2.4. It was instituted at the early stages of research in this domain. It comprises three layers, namely the layers of application, network and perception.

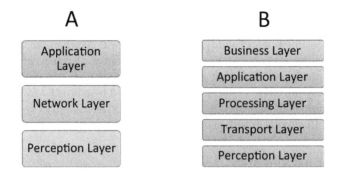

FIGURE 2.4 Architecture of IoT (A: three layers) (B: five layers)

a. The **perception layer** comprises sensors to sense and collect environmental data. This detects certain physical parameters in the world or recognizes certain smart objects.
b. The **network layer** is responsible for linkups to additional servers, network devices and intelligent things. Its characteristics are, in addition, utilized to transmit and process sensor data.
c. It is the responsibility of the **application layer** to provide the user with specific application services. This describes various applications in which, for example, smart houses, smart cities, etc. can be implemented on the Internet of Things.

The design with this trio of layers expresses the primary notion of IoT, but it's not appropriate for IoT research, as that is often based on finer aspects of the IoT. This led to a more complex architectures projected in the literature. One of the designs is of five layers that also comprise the business and processing layers [13] [14] [15] [16]. The five levels are business, application, processing, transport and perception layers (Figure 2.3). The application and perception layer have the same function as three-layered architecture. The remaining layers are illustrated as follows:

i. **Transport layer** transfers data from sensors to and fro between processing layer and perception layer through networks such as NFC, RFID, Bluetooth, 3G, LAN, Wireless, etc.
ii. The **processing layer** is furthermore classified having the status of **middleware**. This collects, analyzes and manages huge data volumes coming through transport layer. Middleware is able to handle the lower layers and support them with varied services. For processing big data, it employs numerous technologies, such as cloud computing, databases etc.
iii. The **business layer** controls the complete IoT network, incorporating apps, market and benefit structures, and the consumers' privacy.

The design anticipated by Wang [17] was influenced as a result of study of information processing within human brain. Human acumen and capability to perceive, sound, recall, take judgment and adjust to their environment has direct influence on their work. Their model comprises three sections:

a. A unit for computing and data management or the data center, akin to the human brain,
b. The elements and sensors in the networking resembling the nerve network.
c. The remote data-processing nodes and smart gateways network similar to the spinal cord.

2.2.6 ECOSYSTEM OF IoT

IoT can be thought of as an ecosystem that provides a data transfer network interconnected with cloud computing and Big Data to offer intelligence for recognizing behaviors and explaining reactions established on data summarized through smart devices/objects accessible in up-and-coming smarter cities devoid of user-to-user communications. Figure 2.5 illustrates the IoT network design, wherein information obtained through smarter cities is integrated into the cloud servers. The contact between cloud and users that are starting to be more involved (presumes) is provided thorough this flow [18]. Centralizing data on behalf of every and all sensors and entities are responsibility of cloud computing system. By formation of an omnipresent network, it also helps in connecting and communicating. For reaching a consensus that permits determination of patterns of human dynamics, the cloud allows for the incorporation of Big Data analysis. Eventually, the human evolutionary structure includes input channels and tools to encourage behavioral change [18] [19].

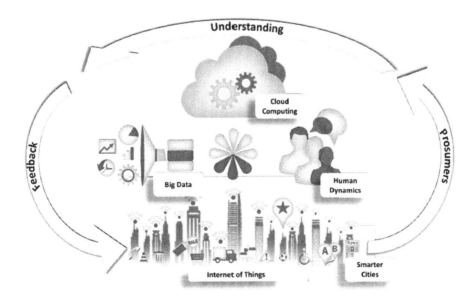

FIGURE 2.5 IoT Ecosystem

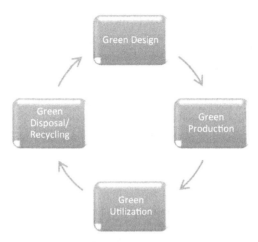

FIGURE 2.6   Life process of Green IoT

2.3   GREEN IOT

Internet of Things (IoT) includes massive usage of the development network and interconnected devices. Therefore, the resources needed to implement the entire element set of networks and their operational energy consumption needs to be kept minimal. Power usage is a crucial and imperative criterion to achieve sustainability in green IoT and smart planet deployment. To have a healthy digital environment, minimal CO_2 pollution and adverse effects on environment, IoT requires improving efficiency of each and every system component, like sensors, computers, software, and utilities.

Figure 2.6 shows the life process of green IoT taking in account green design, green production, green utilization and eventually disposal and recycling to have minimum or no environmental impact [20].

2.3.1   KEY TECHNOLOGIES OF GREEN IoT

The key contribution for making efficient Green IoT is as depicted in Figure 2.7:

i.   **Green Tags Networks**
     The tags used to identify individual items should be environment friendly, maintenance free and sustainable.

ii. **Green Internet Technologies**
     For green internet technology, hardware and software consideration should be taken into consideration where the hardware solution manufactures devices that consume less energy without a reduction of the performance. On the other hand, the software solutions offer efficient designs that consume less energy by minimum utilization of the resources.

iii. **Green Sensing**
     A green wireless sensor network (WSN) is another key technology to enable green IoT. Wireless sensor network (WSN) contains a tremendous number of

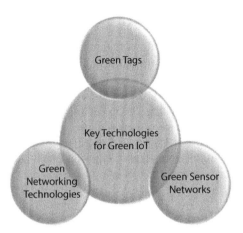

FIGURE 2.7   Key technologies for Green IoT

sensor nodes with limited power and storage capacity. To achieve green WSN, different techniques should be considered

2.3.2 APPLICATION OF GREEN IoT

The numerous Green IoT implementations are as follows:

a. **Smart home**: A G-IoT allows a smartphone or computer to manage illumination, cooking and electronic devices to be controlled remotely. Smart-beds, garbage disposal, lighting fixtures, machines recommending clothes depending on your preference, temperature, doors and walls that can enable unlimited levels of sunshine, digital soundproof rooms, and doors are few more implementations for a smart home. Hidden electronics microphones, sensors, and computers inside houses are soundproof energy fields that you can walk through. A central computer acknowledges voice commands, differentiates inhabitants for custom answers and acts, and the TV, screen and mobile combine in one unit, etc.

b. **Industrial automation:** Industries have been automated with machines that, without or with little manual intervention, allow for fully automated tasks. An internet-based automation network for the market enables the operation of manufacturing equipment by a single industry operator.

c. **Smart healthcare:** IoT intends to remodel the healthcare industry by creating new and sophisticated devices connected to the internet that produce critical real-time data. This helps achieve three main results of effective health services-enhanced accessibility, improved quality, lowered costs.

d. **Smart grid:** In conjunction with IoT, a smart grid concerns efficiency and equilibrium It's about constantly modifying and re-adjusting to produce electricity optimally at the lowest possible expense and the maximum permissible efficiency.

2.4 IOT BIG 7: IOT—CHALLENGES AND PROBLEMS

IoT permits devices/things/objects for communicating data and provides these services via internet protocols. Together with several heterogeneous artifacts, the complexity of the interconnected world demands resolution for seven most important challenges: interoperability, scalability, reliability, efficiency, availability, storage and security. Gartner [21] speculated that, the IoT might comprise 26 billion devices by 2020, creating new problems in all areas pertaining to data centers. IoT technologies often necessitates real-time Big Data processing, resulting in an escalation of data center burden and creating new protection, efficiency, and data analytics challenges [22]. Below is a summary of the difficulties.

2.4.1 IoT Scalability

A scalable IoT system should be capable of connecting new objects/devices, new members, new diagnostic abilities and technology able of imparting continuing support. In case new users or/and devices wish to join the infrastructure or network, IoT scalability ought to regard prospects of offering high QoS (low latency, analytics, etc.). IoT scalability poses a key challenge for our society today, owing to quick escalation of connected devices, numerous technical modifications, humongous IoT interactions, on top of exploiting cloud computing to attend to challenges posed by the Internet of Things as the growing demand for services.

2.4.2 Interoperability of IoT

Despite the complexity of multiple IoT-integrated systems, interoperability is an extra main test that IoT will countenance in order to deliver services effectively and to share data. As per a McKinsey report [23], despite the rapid development of IoT systems, their interoperability is expensive, in the range of four trillion dollars or 40 percent of the total operational value by 2015. Effective IoT interoperability includes specifications that allow different types of devices to connect function and integrate. Acquisition, data sharing, processing, and use of network data have been very problematic so far. Interoperability might be evaluated in the context of considerations like information processing degree (organizational, technical, syntactic and semantic) or at the time of achieving interoperability (dynamic, static) [24] [25].

In the analysis of information there are four directions:

- Organizational (total information relating to organization)
- Technical (means of data representation on physical media)
- Syntactic (syntactic information representation in data gathering, communication, recording, and processing)
- Semantic (the meaning of the data).

The analysis of interoperability is done at the level of software/ hardware, platforms and systems that facilitate device-to-device communication (M2M) at the technical

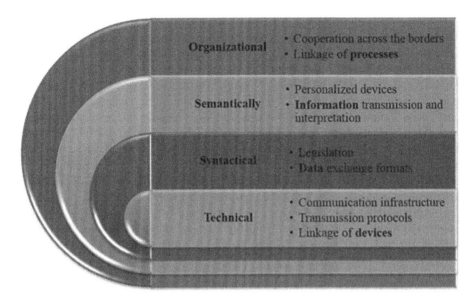

FIGURE 2.8 Challenges by level of interoperability in IoT

level. These investigative analyses concentrate on communication infrastructure and protocols. Starting on the four interoperability levels, Figure 2.8 identifies the interoperability problems that are specific to each level (processes, details, documents, and devices). The IoT constitutes a broad array of applications that pose problems in terms of static interoperability (conformity of requests accomplishment). Partial non-interoperability is acknowledged (e.g., for some protocols), in case they are addressed in sight (dynamic interoperability). Smart middleware and gateways from heterogeneous and dynamic IoT environments have demonstrated complex interoperability capabilities [24].

2.4.3 IoT Reliability

For any device, reliability is characterized by the function of a system's capability for reliably executing any mission/ request/task exclusive of fault/failure, a concept that moreover favors the IoT context. As IoT constitutes large data volumes where information desires proper collection, storage, and control, modern technologies are required to efficiently and effectively disseminate and process information. Many IoT implementations have to run for a limited duration that needs a long-tenure investment. Underneath these scenarios, a network must be adequately adjustable to adapt to environmental circumstances or necessary improvements in components of the network. To ensure better reliability, attempts for standardization are crucial. Kempf et al. (2011) outline four main areas of investigation and analysis that standardization raises: reliability in device architecture design, reliability in system development, load bearing of mobile network sensor gateway connectivity and reliability of communication [26].

Device reliability becomes a real challenge due to the quantity and variety of interconnected objects. The network should consistently provide right, constant and efficient data access, and good data processing depending on the type and function of a computer, even in an instance of missing control, low or no Wi-Fi signal, if an access point or a server breaks down. Network operators are challenged by numerous users, having constant internet connection in support of an assortment of online services, conjointly with IoT network having large number of connected devices, because they must assure continuous and reliable high-speed access to a network. IoT's progress is highly dependent on network efficiency; therefore, the entire IoT infrastructure as well as network providers at each level must be evaluated robustly. An additional challenge in securing uninterrupted broad access at level of complete IoT infrastructure is multiple types of networks or network operators providing several services. The provided service reliability is associated by means of assuring superior availability, compilation, storing and analysis of vast data volumes obtained through IoT devices without mistake. As per these deliberations, IoT interoperability seems strongly linked to reliability.

2.4.4 IoT Efficiency

Another challenge is ensuring efficiency of IoT infrastructure in a connected environment. Network must be capable of:

- Supporting multiple real-time analysis of a vast data volumes existing within network,
- Fulfilling the variety of demands for data processing, and
- Processing data as soon as it enters independent of storage position (application decentralization).

An additional challenge is ensuring advanced data analysis and processing in support of view to of intelligent devices with embedded processes of machine learning. For this reason, explicit constituents like artificial intelligence, intelligent agents, neural networks, etc., should be added to the IoT structure. From the same performance viewpoint, we should look at the IoT network's economic angle and calculate the impact of its activity against the effort put into its operation in a well-defined sense. In fact, it transforms into an aspect of the higher-order feedback loop demonstrating a systematic activity in relation to its setting by giving the IoT a cybernetic paradigm. A systemic approach to the IoT automatically endows it with another efficiency perspective.

2.4.5 IoT Availability

In terms of quality/access, many factors are imperative [15]–[20]: period, place, delivery of service, network, entity and device. Figure 2.9 describes the criteria functionality as well as the challenges faced by those specifications. Augmenting the six factors IoT availability pointed out previously, the environmental facets that are increasingly emerging from latest development of technology, like green cloud, green

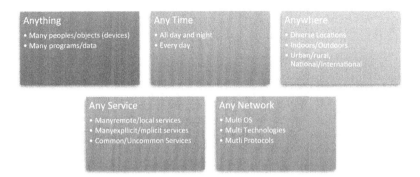

FIGURE 2.9 Availability challenges in cloud computing

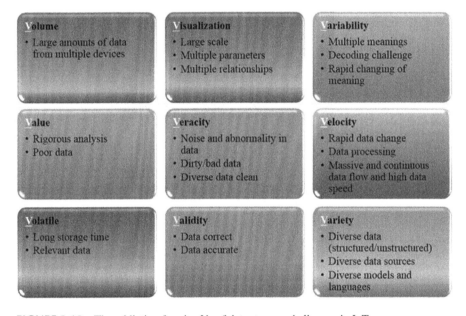

FIGURE 2.10 The addiction for nine Vs of data storage challenges in IoT

computing, including green IoT must also be considered. This green feature is primarily based on the topic of exhaustible resources and at the same time renewable energy sources.

2.4.6 IoT Software

IoT produces vast data volumes that need processing and analysis in real time, which necessitates a large number of activities in data centers, resulting in new security, network capacity, and analytical capability challenges. In addition, the heterogeneity of numerous devices in combination with large data volumes produces challenges for

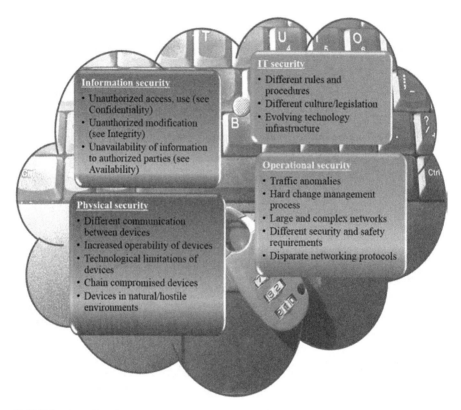

FIGURE 2.11   Security challenges in IoT

data storage management. These challenges related to IoT data include the nine Vs of Big Data: value, volatile, validity, veracity, visualization, variability, velocity, variety, and volume

2.4.7   IoT Security

The IoT presupposes the presence of several connected devices, resulting in several access spots and consequently enhanced safety hazards. However, multiple software standards, middleware convergence, APIs, machine-to-machine connectivity, and so on unavoidably lead to elevated safety threats and an intricate situation [28].

IoT security is equally a concern and a peak priority. We have to guarantee device safety and service security to customers. IoT Security analysis has several dimensions incorporating operational security, physical security, IT security, and information security [17]–[29]. Figure 2.11 exemplifies few challenges to IoT security. The primary purpose of information security is to protect the secrecy, fairness and quality of data. Another important security factor in the IoT sense is protection for non-repudiation functionality of data. This topic remains critical, beyond the four dimensions in IoT data security analysis. In IoT architectures and frameworks, apprehension for quality assurance and standardization development is necessitated due to security concerns. Safety often renders the biggest hurdle along the path of the government-run endeavor for expansion of IoT facilities.

The tenaciously diverse disposition of IoT sections leads to various challenges in IoT security. Contingent upon these facts, IoT might move each and every one flaws of the virtual world in the physical world. Bearing in mind major challenges and challenges faced by the connected environment, experts are continuously searching for resolutions.

2.5 INCORPORATING IOT AND CLOUD COMPUTING FOR ADDRESSING INTRINSIC CHALLENGES OF IOT

The services proffered through cloud computing convey significant advantages by remodeling storage, analytics, resources, computing, network-adjusted systems and services management, control and coordination. Service providers provide IoT advantages on stage of every cloud category: IaaS enables network and device infrastructure management; PaaS supports application environment and OS management; SaaS will handle anything connected to users, even applications [30]. Cloud computing in addition provides the IoT with the opportunity of controlling resource admittance through IaaS services; providing data access through PaaS services; or full admittance toward software applications through SaaS services [31]. IoT solutions delivered by SaaS are developed upon PaaS architecture, allowing business processes through IoT tools and apps to be carried out.

The IoT will allow the most of the three models of cloud computing: private cloud, public cloud and hybrid cloud. For IoT usage, cloud models to be utilized are selected depending upon explicit necessities and safety. Two convergent methods can be adopted by the joint usage of IoT and cloud computing [32] introducing IoT technology into the web (cloud-centric IoT) or using applications in combination with IoT (IoT-centric). Figures 2.12 and 2.13 outline a set of methods for combining IoT with cloud computing along with categories of cloud services and models. The amalgamation of cloud computing and IoT translates into an exemplary literature investigation, for instance Cloud-IoT, Cloud-based IoT, or Cloud of Things.

2.5.1 SCALABILITY VIA CLOUD COMPUTING

An imperative feature assimilated by cloud computing is its versatility in responding to users increasing/diminishing requirements. A function like this allows IoT costs to go down, so consumers pay only for services they use. Cloud enables multi-level scalability, such as network management, existing devices, data volume and data storage facility, data diversity, and applications-related services (horizontal and vertical flexibility).

A major advantage offered by cloud computing is on-demand scalability. While we scrutinize an array of advantages accessible by cloud computing, the power of on-demand scalability is often difficult to conceptualize. Nonetheless, organizations benefit from massive advantages when implementing automated scalability

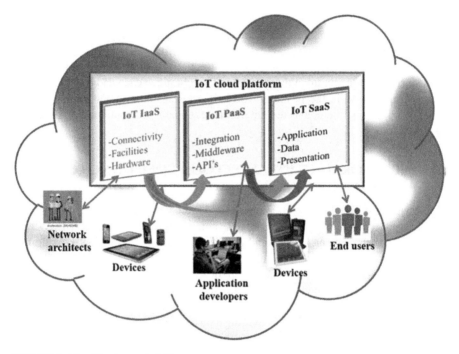

FIGURE 2.12   Cloud-centric IoT

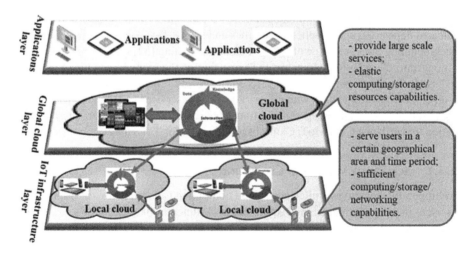

FIGURE 2.13   IoT-centric cloud computing

appropriately. It is clear that the scalability advantages fall in compliance with their underlying complexity within the IoT sense. For example, on-demand scalability of only certain apps needs scalability capability across the whole cloud environment (e.g., traffic delivery in multiple instances).

2.5.2 INTEROPERABILITY THROUGH CLOUD COMPUTING

IoT comprises an assortment of internet-connected objects like 2G/3G/4G, Bluetooth, Wi-Fi, NFC, ZigBee, Z-Wave, and WSAN. In addition, numerous devices functioning over a lone channel creates interaction complexity. An alternative is building hubs, ensuring that they can connect through multiple conduits and accumulate signals as of a wide array of devices [21]–[33]. Cloud computing extends customary interfaces and diverse computer portability across the cloud providers [34].

SaaS providers allow IoT clients to use software directly (over the internet), from anywhere and devoid of requirement for deploying clandestine servers. It provides hardware autonomy by virtualization, which mitigates device dependence on specific hardware. PaaS provides interoperable architectures and middleware for services and data sharing amid diverse devices. Such capabilities for IoT interoperability have recently been developed by services like container-as-a-service and metal-as-a-service.

2.5.3 FUNCTIONALITY VIA CLOUD INFRASTRUCTURE

Another way in which the cloud provides solutions for enhancing system functionality is by increasing the battery life of devices (e.g., through removing the heavy tasks assigned to computers) or by developing a scalable architecture [35] [36]. System efficiency is also enhanced, because cloud computing provides a disruptive-tolerant technology by improved site-redundant cloud services availability [34]. Cloud computing often utilizes assorted procedures for ensuring data synchronization that enhances reliability and accuracy in transactions. Good traffic control can increase network efficiency. Cloud computing provides management procedures that are able to handle unnecessary data transfers and track behaviors for triggering new traffic-sharing instances [37].

2.5.4 EFFICIENCY THROUGH CLOUD COMPUTING

Cloud computing extends various benefits that result in enhanced IoT competence, such as multi-level control, enabling availability management, performance, scalability, better energy usage [34], and on-demand unlimited processing capabilities [35] [39].

2.5.5 AVAILABILITY

Data stored in cloud servers is handled uniformly by benchmark API interfaces [40], and is capable of accessibility and processing from anywhere [41]. The cloud environment provides efficient outcomes for connecting, tracking, and managing anything/object (devices), regardless of time and place, by means of embedded applications and custom gateways [41].

2.6 SUMMARY AND CONCLUSIONS

The proliferation of devices with imparting impelling technologies puts the dream of an Internet of Things back together, where tracking and incitement functions regularly blend out from plain sight and new capabilities are rendered feasible through leveraging rich new data sources. When developing new applications, the creation of the cutting-edge portable platform must depend on the inventiveness of the clients. IoT is a great invention that will affect this area by providing new knowledge and the technological tools needed to make innovative applications.

The IoT problem linked to increasing data storage can be addressed by cloud storage, as it extends with flexibility and protection, and is capable of customization to IoT network requirements. Cloud computing proffers limitless, inexpensive methods for storing and processing organized as well as unstructured data by implicit storage schemes. Cloud data can be covered by high-level protection implementation [38]. Nevertheless, cloud computing security is even now a significant contest that can proliferate even in the direction of IoT. As pointed out, protection impediments for both IoT and cloud computing are obstacles for growth and widespread acceptance of the two paradigms in more critical fields such as industry.

REFERENCES

[1] S. Shahrestani, H. Cheung, and M. Elkhodr, "The Internet of Things: Vision & Challenges," in *2013 IEEE Tencon*, 2013, pp. 2018–222.

[2] Accenture Strategy, "SMARTer2030: ICT solutions for 21st century challenges," Global eSustainability Initiative (GeSI), Brussels, Belgium Technical Report, 2015.

[3] Green Power for Mobile, "The Global Telecom Tower ESCO Market," Technical Report 2015.

[4] G. Fettweis, J. Malmodin, G. Biczok, and A. Fehske, "The global footprint of mobile communications: The ecological and economic perspective," *IEEE Communication Magazine*, 2011.

[5] IMT Vision-Framework and Overall Objectives of the Future Development of IMT for 2020 and Beyond, 2015, Document Rec. ITU-R M.2083-0.

[6] M. Albreem, "5G wireless communication systems: vision and challenges," in *2015 IEEE International Conference on Computer, Communication and Control Technology, Malayia*, 2015.

[7] www.ida.gov.sg/~/media/Files/Infocomm%20Landscape/Technology/Technology Roadmap/InternetOfThings.pdf

[8] Norway, Dr. Peter Friess EU, Belgium Dr. Ovidiu Vermesan SINTEF, "Internet of Things: converging technologies for smart environments and integrated ecosystems," in *River Publishers' Series in Communications*, River Publishers, 2013.

[9] Norway, Dr. Peter Friess EU, Belgium Dr. Ovidiu Vermesan SINTEF, "Internet of Things: from research and innovation to market deployment," in *River Publishers Series in Communications*, Aalborg: River Publishers, 2014.

[10] https://dzone.com/articles/the-internet-of-things-gateways-and-next-generation.

[11] http://tblocks.com/internet-of-things

[12] www.reloade.com/blog/2013/12/6characteristics-within-internet-things-iot.php.

[13] O. Alsaryrah, T.-Y. Chung, C.-Z. Yang, W.-H. Kuo, and D. P. Agrawal I. Mashal, "Choices for interaction with things on Internet and underlying issues," in *Ad Hoc Networks*, 2015, pp. 68–90.

[14]   O. Said and M. Masud, "Towards internet of things: survey and future vision," in *International Journal of Computer Networks*, vol 5, 2013, pp. 1–17.

[15]   T.-J. Lu, F.-Y. Ling, J. Sun, and H.-Y. Du M. Wu, "Research on the architecture of internet of things," in *3rd International Conference on Advanced Computer Theory and Engineering (ICACTE '10), vol. 5*, Chengdu, China, August 2010., pp. 484–487.

[16]   S. U. Khan, R. Zaheer, and S. Khan, R. Khan, "Future internet: the internet of things architecture, possible applications and key challenges," in *10th International Conference on Frontiers of Information Technology (FIT '12)*, December 2012, pp. 257–260.

[17]   H. Ning and Z. Wang, "Future internet of things architecture: like mankind neural system or social organization framework?" in *IEEE Communications Letters*, 2011, pp. 461–463.

[18]   Y. Bocchi, D. Genoud, and A. Jara, "The potential of the internet of things for defining human behaviours," In *International Conference On Intelligent NETWORKING And Collaborative Systems*, 2014, Pp. 581–584.

[19]   S. Chen, S. Xiang, Y. Hu, and L. Zheng, "Research of architecture and application of internet of things for smart grid," in *International Conference on Computer Science and Service System*, 2012, pp. 938-939.

[20]   S. Murugesan, "Harnessing green IT: Principles and practices," in *IEEE IT Prof.*, 2008, pp. 24–33.

[21]   Gartner, "The impact of the internet of things on data centers. Gartner report," 2014.

[22]   R. Davis, "Big problems with the internet of things," 2014. www.cmswire.com/cms/internet-of-things-7-big-problems-with -the-internet-of-things-0214571.php

[23]   Iconectiv, *Overcoming interoperability challenges in the Internet of Things*. Telcordia Technologies, 2016.

[24]   IERC, Internet of things IoT semantic interoperability: Research challenges, best practices, recommendations and next steps, IERC, 2015.

[25]   A. Wiles and H. Van der Veer, "Achieving technical interoperability ," *The ETSI – approach ETSI* 3rd edition, April 2008, 2008.

[26]   J. Arkko, N. Beheshti, K. Yedavalli, and J. Kempf, "Thoughts on reliability in the Internet of Things," in *Proceedings of the Interconnecting Smart Objects with the Internet Workshop*, Prague, Czech, 25–26 March 2011; pp. 1–4.

[27]   B.A. Bagula, "Internet-of-things and big data: Promises and challenges for the developing world," 2016. http://unctad.org/meetings/en/Presentation/ecn162016p16_Bagula UWC_en.pdf.

[28]   C. Kocher, "The internet of things: challenges and opportunities," 2014. Sand Hill. http://sandhill.com/article/the-internet-of-things- challenges-and-opportunities/

[29]   R. Liwei, "IoT security: problems, challenges and solution," 2015. www.snia.org/sites/default/files/DSS-Summit-2015/presentations/

[30]   Pasquier T, Bacon J, Ko H, Eyers D Singh J, "Twenty security considerations for cloud-supported internet of things," in *Internet Things*, 2016, pp. 269–284.

[31]   J. Soldatos, "IoT tutorial: chapter 4 – internet of things in the clouds," 2016. www.linkedin.com/pulse/iot-tutorial-chapter-4-internet-things-clouds-john-soldatos

[32]   IoT tutorial: chapter 4 – internet of things in the Clouds, "IoT – Internet of Things," , 2016. www.linkedin.com/pulse/iot-tutorial-chapter-4-internet-things-clouds-john-soldatos

[33]   Workflow Studios (2016) "Taming the Internet of Things with the cloud," March 22, 2016. https://workflowstudios.com/taming-the-internet-of-things-with-the-cloud/

[34]   I. Llorente, "Key challenges in cloud computing to enable future internet of things," in *The 4th EU-Japan Symposium on New Generation Networks and Future Internet*, 2012.

[35]   B. Yuxin and M. Yun, "Research on the architecture and key technology of Internet of Things (IoT) applied on smart grid," in *Advances in energy engineering (ICAEE), International Conference on Advances in Energy Engineering, Beijing*, 2010, pp. 69–72.

[36]   Z. Bi, LD Xu, and C. Wang, "IoT and cloud computing in automation of assembly modeling systems," in *IEEE Trans Ind Inf 10(2)*, pp. 1426–1434.

[37]   R. Adams and E. Bauer, "Reliable cloud computing – key considerations," 2017. www.nokia.com/blog/reliable-cloud-computing-key-considerations/

[38]   S. Mohapatra, P.K. Pattnaik, and S.K. Dash, "A survey on application of wireless sensor network using Cloud Computing," *Int J Comput Sci Eng Technol*, pp. 50–55.

[39]   P. Parwekar, "From Internet of Things towards cloud of things," in *Computer and Communication Technology (ICCCT), 2011 2nd International Conference on Computer and Communication Technology*, 2011, pp. 329–333.

[40]   S. Kamburugamuve, R.D. Hartman, and G.C. Fox, "Architecture and measured characteristics of a cloud based internet of things," in *Collaboration Technologies and Systems (CTS), 2012 International Conference on Collaboration Technologies and Systems*, 2012, pp. 6–12.

[41]   P. Saluia, N. Sharma, A. Mittal, and S.V. Sharma, and B.P. Rao, "Cloud computing for internet of things & sensing based applications." in *Sensing Technology (ICST), 2012 Sixth International Conference on Sensing Technology*, 2012, pp. 374–380.

3 Green-IoT (G-IoT) Architectures and Their Applications in the Smart City

Mayank Goel, Amber Hayat, Akhtar Husain, and Surjeet Dalal

CONTENTS

3.1 INTRODUCTION

Technology, which is making changes in every area and automating things for the ease of life, produces the best practice to grow and learn. It works towards the betterment of humanity and providing various helping features for people. Technical things are all controlled by humans which helps the general public to understand the future technology and adopt it in our lives. This will help our quality of living and provide a different exposure to learn and work. We all want the futuristic design and automation of daily work. Kevin Ashton came up with an idea of working with a technology named as Internet of Things (IoT), which generally means bringing all objectives under one roof with the help of the internet, which provides the interconnectivity and a working platform for its operations. Now IoT is emerging everywhere, people and companies are working hard to connect of all the appliances and for various gadgets to be controlled from remote locations for more convenience. This technology will definitely help people to live their life with more comfort and understand the demands to upgrade to new tech.

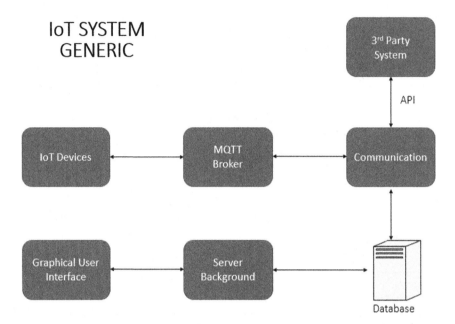

FIGURE 3.1 The different systems used under the working of IoT as technology

3.2 WHAT IS IOT?

Internet of Things (IoT) links everything together, and lively utilization of IoT innovation is an alluring field of exploration. IoT proposes to provide the basic working of all the technology under one control node. IoT tells the public that they can have control over their objectives and provides them with many new features. IoT has inspired a lot of new technologies which will be used in the coming decade. Technology will come close to humans, and technology can help the automation of different things and their remote control.

IoT is a promising and growing technology. It allows people and things to be connected anywhere, anytime, with anyone and anything, using any link and any service. Generally, IoT is a platform of different sensors and various technologies are embedded in it. This allows IoT to grow and stand as an innovation which will be everywhere in the coming years. It helps the humans to connect with every aspect of life. The IoT sensors and devices collect all types of data and then store that info which will further be analyzed to get more useful information. It helps humans to develop along with the futuristic idea and technology.

There are various aspects of IoT technology which are listed below:

1. Connectivity: Internet plays a role in connectivity between the various resources and providing the best of services to control the different devices involved in its working. It helps in web services, IP connection, and cloud computing.
2. Working Node: The middleware technologies will help for data storage, processing, computing, finding resources.

3. Hardware: The hardware plays a major role in finding the best of services by collecting the data and then sending them for information to be extracted from them. This includes the various sensors, tags, devices, actuators.
4. Interface: It provides the pathway from the working node to the customer connected to it. This will help them to visualize the things and can take part in the process.

However, the research into the IoT is still in its infancy and there are many key challenges needed to be addressed such as the battery life, simplicity data context-awareness, privacy and security concerns, multiple active things, interference-free connectivity, the cost of terminal devices, scalability, and heterogeneous terminal devices.

The Internet of Things (IoT) is an ecosystem which is not only a network to transfer data, but also interconnected with big data and cloud computing to provide intelligence, in order to be able to recognize the behaviors, and even explain actions according to the information captured by the smart objects that are available around the emerging smarter cities without requiring human-to-human or human-to-computer interaction.

3.3 GREEN IOT

The IoT is doing an enormous amount for the betterment of the technology and helping humans to learn and incorporate the latest developments in their daily life. IoT is doing wonderful job, but a question arises, what about the wastage of resources involved in all this? This question is a big issue because of e-waste, the type of materials used in manufacturing, and the costs incurred. This all makes IoT inefficient, which has led to the idea for a Green Internet of Things (G-IoT). What is G-IoT? How can it help? What is the difference between G-IoT and IoT? Is it beneficial? Is it cost efficient? These all questions are very big to deal with, but first we need to understand the basics and then we should look at the different issues in this new technology.

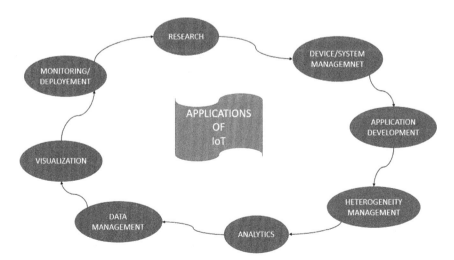

FIGURE 3.2 These are the Applications of IoT, which are mainly used and worked on

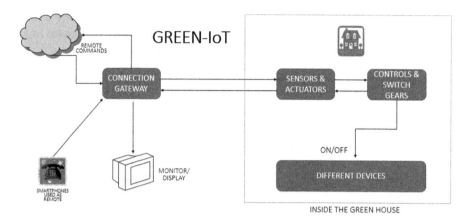

FIGURE 3.3   The architecture of Green IoT and its working

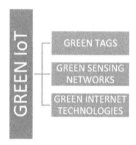

FIGURE 3.4   Green IoT and its main objectives

Inspired the accomplishment of a low control utilization IoT, Green IoT is proposed. Because of quick headway in computerized innovation, there has been tremendous change in practically every area of work. There has been an expanding pattern of utilization of brilliant gadgets. G-IoT is anticipated to bring critical changes in our day-to-day life and we need to understand the vision of "green encompassing insight."* Within a couple of years, we will be surrounded by a huge measure of sensors, gadgets, and "things," which will have the option to convey information by means of 5G, act "brilliantly," and give green help to clients in dealing with their assignments.

Because of the developing familiarity with ecological issues far and wide, Green IoT innovation activities are a priority. Greening IoT alludes to the innovations that make the IoT natural in a neighborly manner by empowering users to store, get to, and oversee different data. Green IoT is the term utilized for aggregate advancements of IoT where parts get used and reused number of times.

As this idea inspires new tech and new peripherals which can work in it, this all can impress and encourage the world to use it. Here are some ways to address the ecological impacts while utilizing IoT.

1.   Green Use: Reducing the actual utilization of PCs and other data frameworks as much as utilizing them in an ecologically stable way.

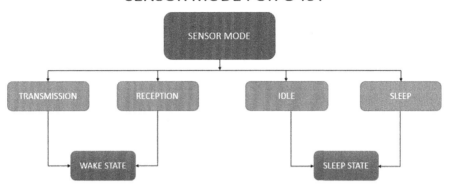

FIGURE 3.5 How the sensor performs its task under the IoT work flow

2. Green Disposal: Refurbishing and reusing old personal computers and reusing unwanted PCs and other electronic gear.
3. Green Design: Designing effective and ecologically stable segments, personal computers, and servers and cooling equipment.
4. Green Manufacturing: Manufacturing electronic parts, personal computers, and other related sub frameworks with negligible effect or no effects on the earth.

These are some of the areas which will be helpful for the growing and emerging of G-IoT in the world and help the public.

3.4   A BETTER IoT

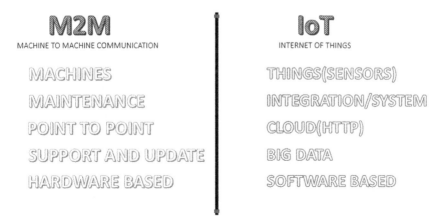

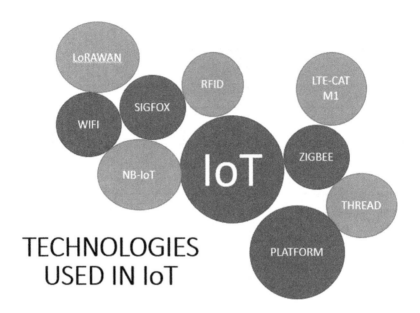

TECHNOLOGIES
USED IN IoT

3.5   PROBLEMS/CHALLENGES/ISSUES—G-IOT

There are numerous inventions in process to achieve a better future. A future is being developed, enabled by green technology. G-IoT is still in its infancy. This means the scientist are working really hard, but need time to develop it completely for it to be useful for the people around. The different challenges are software based, hardware based, internet protocol connection, policy working. The various factors which are causing challenges are listed here.

1. Integration of all the devices to work as to provide maximize energy efficiency for the customer and care for the nature.
2. Devices and sensors to be made environmentally compatible, so as not to disturb the environment in its working.
3. Various energy conservations models to be made and implemented in each program around its working.
4. The power consumption and energy efficiency to be maintained between the communication of sensors, devices, protocols, etc.
5. The reduction of complexity of the system in the Green IoT infrastructure.
6. Cloud management system to be less complex and more efficient.
7. Efficiency of the time and controlling node to be taken care of for various security activities like of encryption, decryption, data security, various algorithms.

While discussing these points regarding the efficiency of the objects in the Green IoT, we need to consider the research and innovation fields to know better the new technology which is to be introduced in the near future. The research fields which need to develop optimal and efficient solutions for greening IoT are:

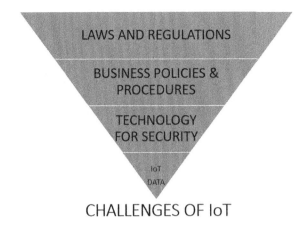

FIGURE 3.6   Challenges incurred while performing IoT tasks

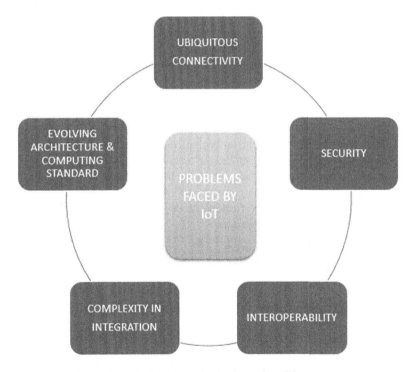

FIGURE 3.7   Problems faced by IoT execution in the real world

1. Unmanned aerial vehicles (drones) need to come up and replace lots of IoT devices. This can help IoT to make a difference. The different fields in which it can be used are agriculture, traffic, surveillance. This will bring Green IoT at great efficiency and minimal cost.

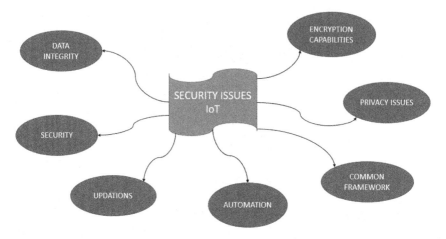

FIGURE 3.8   Issues of security under IoT

2. New separate cloud services are to be launched, which will help the IoT plat-
    form to be independent and more secure around the other resources. The
    sensors and cloud integrate using a common session they create after every
    instance of data processing. A green social network as a service (SNaaS) is
    to be launched, which can be used for energy efficiency services, network
    storage, and various cloud activities.
3. Machine-to-machine communication has to be taken care of and people are
    going to automate the machines by implementing it. This technology reduces
    the work force needed and makes the machines self-working.
4. IoT devices are very small in size, so big batteries can't be installed on those
    devices. Due to the internet connection they are taking a lot of battery usage.
    Thus, many companies are working on devices which have low battery con-
    sumption in their working process.
5. The most important work is to reduce CO_2 emissions. This should be con-
    trolled by devices and sensors and this all leads to a smarter and cleaner
    environment.

3.6   METHODS OF G-IOT

Method defines the various aspects which can be controlled and can worked on to
help the public. One way of doing this is to implement environmental protection
using various sensors and devices. This can be very helpful for controlling pollution.
IoT is making life much better and more comfortable. With the help of small-sized
sensors and devices we can track radiation, air quality, water quality, and various natural indicators.

The various factors to look in for the betterment of the public are:

1. Water Management: Water is essential to human life. It is an integral support
    system for humanity. Water makes up approximately 75 percent of the earth's

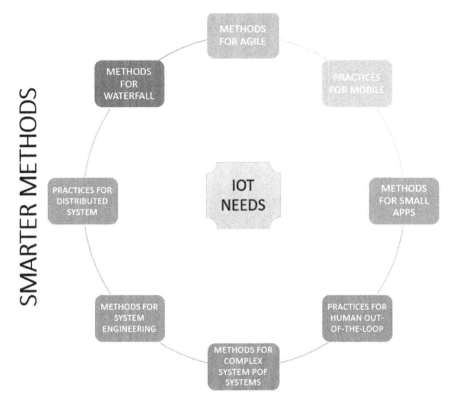

FIGURE 3.9 New practicalities and technologies to be used while performing IoT activities

surface, but drinking water is only 2–3 percent. We need smart water meters to control the water usage.

2. Energy Efficiency: Energy plays a major role at every step. Without the help of energy, we are unable to process the computational work. It helps us to carry out various tasks and perform operations. Energy of any form and factor should be used efficiently. For example, if there is no one in a room, lights should be turned off immediately – this could be a sensor-based without human intervention.

3. Waste Controlling: The waste of any material is a keen issue for the environment. We need to be very careful about discarding waste material which might be used to make newer products for market. Reusability is a great factor which can control the various aspects of product that can be recycled and which will definitely help society.

4. Reduction of Emissions: Products and vehicles are now emitting a lot of noxious gases, which is causing a lot of respiratory and health-related issues. Thus, we should have sensors and anti-emitting gases to control these gases.

COMPONENT WORKING

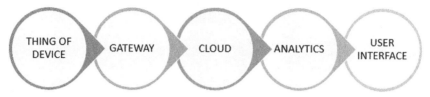

FIGURE 3.10 The five integral components of working in IoT

5. Renewable Energy: Energy conservation is a priority for any government around the world. Various sensors and tags are installed in buildings to help save energy.
6. Adapting Digital Platforms: Digital platforms are very helpful in controlling and monitoring. They get the systems get going and this can also help the customer to connect with technology directly. The digital platforms are easy to operate and very interactive to use with customer care support provided. For example, there is a digital professional networking platform (linkedin) where people around the world can connect with each other professionally.
7. Electric Vehicles: The adoption of electric vehicles is very useful in reducing pollution and using no fuel. But we need the whole system to be connected by the internet and also provide better results for vehicle mileage.
8. Monitoring: Monitoring is the controller node in this system. The controller node should be well designed and controlling power should be implemented in the node. Thus, from remote places we would be able to get information and identify problems and their solution. The monitoring of various devices and sensors can be done easily with no hassle to humans. Future implementation of big data and artificial intelligence can be used for better results and experiences.

3.7 CONTRIBUTION TO WORLD

The IoT is the future business The IoT, creatively, understands complications and then finding its uses to work. The IoT comes in the picture in bringing energy efficiency and reducing damage to the environment.

The data collected by the IoT devices plays a major role because of its importance for the well-being of the surrounding environment. Here are some of the practical contributions of IoT which are very useful for humans:

• E-Waste: E-waste creates a lot of electronic equipment. People are just going on wasting the products by discarding them and not reusing the waste. This is why the reuse of the electronic products is not being carried out.
• Smart Infrastructure: Infrastructure is creating a big issue in life. With some buildings there are various issues with ventilation, heating, cooling, lighting, and security.

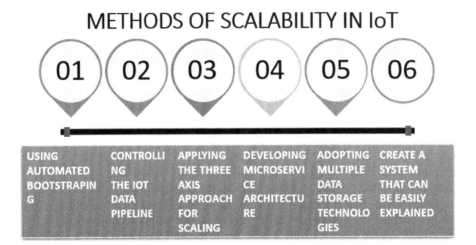

FIGURE 3.11   To enhance the productivity of IoT and its working

- Better Farming: Farmers are the most deprived of technology, and get things done by using traditional methods. The methods that can be used for them are: smart analyzer, soil testing, tested seeds, easy transportation and storage, continuous monitoring, animal detection – all of which can help them to produce efficient crops with less effort.
- Better Air Index: The air quality index is worsening day by day due to pollution which creates issues for people to breathe and to live their life. With the help of G-IoT we can analyze the air quality and modify this by taking various steps which will eventually lead to improvements.

3.8   FUTURE SCOPE OF G-IOT

In the coming decades, we all need to be secure and more innovative. This needs a lot of study and also smart innovations to adopt the new technologies. Ample time and adequate research are needed. We all adapting to the new era by hands-on experience of technology, like that of mobile phones. In the past no mobile phone featured in day to day life, but now almost every person is carrying a smartphone.

IoT as a technology helped to transform our surrounding by making a pathway to interact with machines and carry out tasks. IoT is bringing a digital revolution around the world. This is possible because of the increasing demand and also automating machines to work. This has helped improve the quality of life, the way we work, the way we carry out our day-to-day activities. The system helps us to upgrade our lifestyle.

There is a new concept known as Blue IoT which is also a part of Green IoT. Blue IoT relates to energy consumption and infrastructure services. In this we create a space which is fully energy efficient, equal lighting, safety measures, providing security mechanisms, integrated with smart sensors and devices, which are already installed and a person just needs to come, provide input, and get the work done. They provide the concepts of:

a. Data analytics and IoT
b. Secure and healthy buildings
c. Maintain and work on critical facilities

Due to increased demand to adopt new technology, our Indian government has supported startup businesses by providing them with better facilities, and full support to innovate new things for betterment of the country. This all leads to better understanding of upcoming products and new smart technologies can be better understood by the people. Such startup business is a great asset for any country to come up with smart working on the projects and helps to grow the country. The various startups are working on energy conservation, adoption of electric vehicles, smart farming for farmers – these are some of the hot topics to be discussed and worked on.

3.9 CONCLUSION

The innovators are working hard to bring something new and real systems to understand the demands of people and help them to adapt the working of new technology. These innovators of IoT are understanding the system and with the help of smart sensors, small tags, actuators, combined together make an interface to work on connecting the different things. People are working hard to understand the importance of the new technology and improve it as new projects in IoT, Green IoT, or Blue IoT. This will help conserve resources. Adopting the new technology can be risky at various moments but understanding the concept fully helps to ease the work. The Green IoT replaces IoT, making best usage of biodegradable devices, helping the environment without damaging nature. The Green IoT is the future which needs to be adopted.

BIBLIOGRAPHY

[1] M.A.M. Albreem et al., "Green Internet of Things (IoT): An overview," Proceedings of the 4th IEEE International Conference on Smart Instrumentation, Measurement and Applications (ICSIMA) 28–30 November 2017, Putrajaya, Malaysia.
[2] "What technologies are used in IoT – technology behind Internet of Things." www. avsystem.com/blog/iot-technology/ accessed Feb. 2020.
[3] K. Ashton, "That 'Internet of Things' thing in the real world, things matter more than ideas," *RFID Journal*, 22(7), 2009, 49–59.
[4] Q. Meng and J. Jin," The terminal design of the energy self-sufficiency Internet of Things," 2011 International Conference on Control, Automation and Systems Engineering (CASE), 2011, pp. 1–5.
[5] Keysight Technologies, "Battery Life Challenges in IoT Wireless Sensors and the Implications for Test," Application Note, 2015.
[6] S. Tozlu, M. Senel, W. Mao, and A. Keshavarz Ian, "Wi-Fi enabled sensors for Internet of Things: A practical approach," *IEEE Communications Magazine*, 50(6), 2012 pp. 134–143.

[7] L. Roselli, N. Carvalho, F. Alimenti, P. Notte, G. Orecchini, M. Virili, C. Mariotti, R. Gongalves, and P. Pinho, "Smart surfaces: Large area electronics systems for Internet of Things enabled by energy harvesting," *Proceedings of the IEEE*, 102(11), 2014, pp. 1723–1746.

[8] J. Wu, I. Bisio, C. Gniady, E. Hossain, M. Valla, and H. Li, "Context aware networking and communications: Part 2," *IEEE Communications Magazine*, 52 (8), 2014, pp. 64–65.

[9] Z. Yan, X. Yu, and W. Ding, "Context-aware variable cloud computing," *IEEE Access*, 5, 2017, pp. 2211–2227.

[10] D. Martin, C. Lamsfu, and A. Alzua, "Automatic context data life cycle management framework," International Conference on Pervasive Computing and Applications, 2010.

[11] Z. Yan, X. Yu, and W. Ding, "Context-Aware Verifiable Cloud Computing," *IEEE Access*, vol. 5, 2017, pp. 2211–2228.

[12] M. Ali, S. Khan, and A. Zomaya, "Security and dependability of cloud-assisted Internet of Things," *IEEE Cloud Computing*, 3(2), 2016, pp. 24–26.

[13] N. Simoes and G. Souza, "A low-cost automated data acquisition system for urban sites temperature and humidity monitoring based in Internet of Things," 2016 International Symposium on Instrumentation Systems, Circuits and Transducers, 2016, pp. 107–112.

[14] D. Wu, L. Bao, and C. Liu, "Scalable channel allocation and access scheduling for wireless Internet of Things," *IEEE Sensors Journal*, 13(10), 2013, pp. 3596–3694.

[15] Y. Han, Y. Chen, B. Wang, and K. Liu, "Enabling heterogeneous connectivity in Internet of Things: A time-reversal approach," *IEEE Internet of Things Journal*, 3(6), 2016, pp. 1036–1047.

[16] F. Shaikh, S. Zeadally, and E. Exposito, "Enabling technologies for green Internet of Things," *IEEE Systems Journal*, 11(2), 2017, pp. 983–994.

[17] C. Perera, A. Zaslavsky, P. Christen, and D. Georgakopoulos, "Context aware computing for the Internet of Things: A survey," *IEEE Communications Surveys & Tutorials*, 16(1), 2014, pp. 414–445.

[18] D.-N. Le, B. Seth, and S. Dalal, "A hybrid approach of secret sharing with fragmentation and encryption in cloud environment for securing outsourced medical database: A revolutionary approach," *Journal of Cyber Security Mobility*, 7(4), 2018, pp. 379–408

4 Practical Implications of Green Internet of Things (G-IoT) for Smart Cities

Bijeta Seth, Surjeet Dalal, and Neeraj Dahiya

CONTENTS

4.1 INTRODUCTION

The internet has connected physical objects and communication devices through universal communication via Transmission Control Protocol and Internet Protocol. The internet has changed our lives drastically. Smooth connectivity of the prevalent systems is a vital aspect of the Internet of Things (IoT). It is the IoT technologies that enable machines to work in a smarter, efficient, and effective way by handling computation and communication of data intelligently. IoT has produced innovations in e-healthcare, home mechanization, industry automation, etc. To accomplish smart world improvement and sustainability, Green IoT seeks to diminish carbon emissions and power depletion. Green IoT denotes the IoT environmentally friendly approach, constructing a practice of amenities and storage which enable subscribers to collect, hoard, access, and use information. Smart cities nowadays are considered to be effected most successfully by Green IoT. Now objects in smart cities can interconnect with individuals through IoT. Smart cities will become greener by being able to perceive contamination using ecological sensors. To preserve the sustainability of green habitation in smart cities, Green IoT works in a concerted style. Administrations and societies around the world are doing a lot of hard work to promote the significance of the declination of energy depletion and carbon creation as well as accentuate Green IoT for smart cities.

This chapter presents a brief introduction of enabling technologies for smart cities highlighting their main characteristics, functionality, architecture, and role they play. Section 4.2 describes the motivation and objectives of work. Section 4.3 presents IoT, outlining its traits, applications, opportunities, and barriers to its adoption. Section 4.4 looks at the trairts of Green IoT, lifecycle, barriers to its adoption, and applications. Section 4.5 gives a smart city overview, applications, and technologies. Section 4.6 covers structural design for smart cities.

4.2 INTRODUCTION TO IOT

The Internet of Things is considered to be a new wave of the global economic development in terms of information and industry. The remarkable expansion of expertise has transformed how we work and live. It is the next big thing in the Information and Communication Technology (ICT) industry, capable of connecting small-scale to large-scale systems.

IoT is an extension of and further development of the "Anytime-Anywhere-Anyhow" communication paradigm of ICT. It was first proposed by Kevin Ashton in 1998 [1]. The IoT vision was introduced [2] by the MIT Auto-ID. International Telecommunication Union (ITU) [3] then presented the concept of IoT. It is illustrated in Figure 4.1.

Several countries have put forward long-term nationwide policies for the IoT. Broadband access in Japan[4] is pervasive and adapted to the general public with the aim of achieving communication between individuals, societies, and organizations. Singapore's next generation I-hub [5] aims to provide a safe and universal network. In January 2009, the IBM CEO anticipated the notion of "Smart Earth" in a conference with US president Barack Obama and some business leaders, whereby IoT can be effectively accomplished by mounting sensors in rivers, power grids, railways, roads, trucks, and other objects [6].

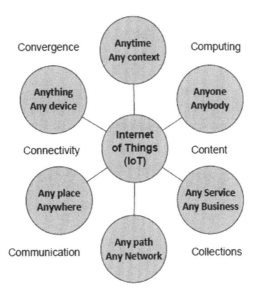

FIGURE 4.1 A high level view of IoT

Components in IoT have been defined as:

1. Hardware: it encompasses the implanted communications hardware such as sensors, actuators, and transceivers.
2. Software
3. Cloud

In general, IoT *hardware* includes:

1. Sensors: they collect and send data to microcontrollers connected to the internet. The sensors consume power and thus they should be designed to be environment friendly. Examples include temperature sensors, proximity sensors, image sensors, motion detection sensors, gas sensors.
2. Microcontrollers
3. Communication: several communication protocols, like Near Field Communication (NFC), Z-wave, Wi-Fi, are available. They are used to connect different things or objects or machines together.

The *elements* in IoT [10] are listed below and depicted in Figure 4.2.

1. Things/devices: act as a sensor which senses its surroundings, for instance, Radio Frequency Identification (RFID).
2. Internet: it provides communication between devices or things with the help of internet protocols.
3. Processing: of data received by sensor or things and generating an output with the help of a micro-controller or micro-processor.

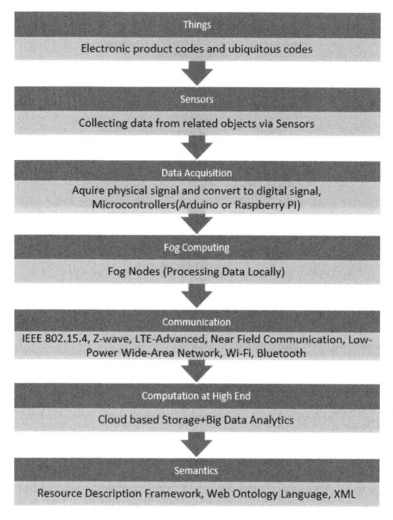

FIGURE 4.2   IoT elements

4.2.1 APPLICATIONS OF IOT

IoT is playing a significant role today. Applications of IoT are discussed below and shown in Figure 4.3 [4, 7, 8, 10].

- Smart Home: IoT has got a lot of applications but smart home is one of the most prominent ones. It has been found that 60,000 people are searching for the keyword "smart home" every month. Almost 256 companies and startups are in the database of the IoT Analytics Company for smart home. More research and companies are focusing on smart home compared to other applications of IoT. Total investment or funding for smart home startups has exceeded $2.5bn. Examples of these startup companies are Nest or AlertMe, Philips, Haier, Belkin.

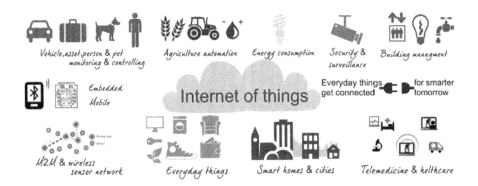

FIGURE 4.3 Applications of IoT

- Wearables: Wearables are also an application of IoT. Examples are Apple smart watch, LookSee bracelet. Wearables maker Jawbone has the biggest funding among all IoT startups.
- Smart City: Smart city poses a lot of challenges to researchers and startup companies have a wide variety of use cases, from water distribution systems to traffic management, waste management, and so on. Its popularity is because the IoT promises solutions to alleviate real problems of metropolitan cities, including traffic congestion problems, reducing noise and pollution.
- Smart Grids: This is a special case which promises to apply analytics based on the information of electricity consumption behavior as well as electricity suppliers and consumers, leading to the improvement of the efficiency, reliability, and economics of electricity. Nearly 41,000 Google searches per month shows the popularity of the concept.
- Industrial Intranet: The industrial internet has also got the applications for IoT. It has not got as much attention as smart home or wearable but still a few market researchers like Gartner or Cisco see the industrial internet as one of the applicatiosn of the IoT leading to highest overall potential.
- Connected Car: This is a little bit slower in research since the development cycle of the automotive industry take two to four years. Large auto makers and other startups have started working on it
- Connected Health (Digital Health/Telehealth/Telemedicine): The medical industry has valuable potential for IoT application. Connected health is a silent giant of the IoT.
- Smart Supply Chain: IoT has been giving a new push to supply chains and making them smarter. It helps in providing solution for taking goods on the road or it helps suppliers in exchanging inventory information.
- Smart Farming: Smart farming is one application of IoT which has large potential but has not been explored much. This technology can change the concept of farming as remote locations of faming can be monitored and livestock can be tracked. It will become one of the important applications in the countries where agriculture is dominant [9].

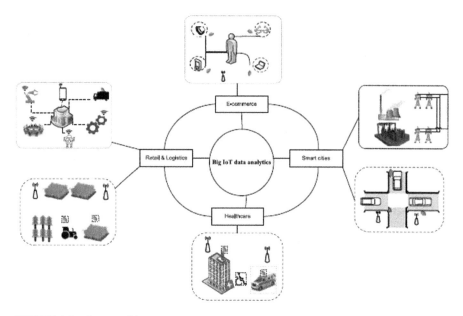

FIGURE 4.4   Opportunities

4.2.2 Opportunities

IoT is presently understood to be one of the most significant transitions in technology. Existing IoT delivers numerous prospects for large information analytics. Figure 4.4 illustrates [23] the instances of use cases and occasions of installing IoT.

- E-commerce: The aggregation of big data and IoT introduces novel issues and occasions to construct a smart milieu. Big data analytics process real-time data and produce outcomes within a time constraint. The prominent areas of analytics are in e-commerce, financial expansion, accuracy of forecast results, improved client separation.
- Smart cities: Efficiency gains can be accomplished by collecting big data through appropriate analytics infrastructure. Different devices are allied to the internet in a smart environment of sharing information. Cloud computing technology has reduced the cost of storage of data. Consequently, the usage of big data in a smart city has the potential to renovate each area of the economy of a nation.
- Marketing and logistics: IoT is reliable in performing crucial tasks in merchandising and logistics to maximize profits.
- Healthcare: Information analytics permits healthcare specialists to identify severe ailments in their early stages to assist in saving lives, safeguards the protection of patients, access physical conditions of the patient, and also increase client satisfaction and retention.

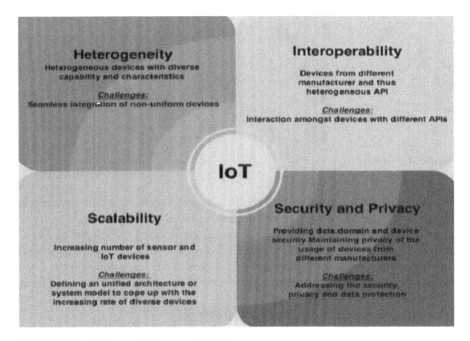

FIGURE 4.5   IoT research issues and challenges

4.2.3 BARRIER TO ENTRY AND FUTURE CHALLENGES OF IoT

While the IoT in today's world has brought extra financial and other benefits, the evolution of the IoT was very nearly stopped in its tracks because of the failure of the earliest market entrants to identify that certain things were indispensable to ensure rapid, safe acceptance of IoT solutions. Certain key issues in IoT have been recognized by researchers as heterogeneity, scalability, interoperability, and security and privacy, [10] and are projected in Figure 4.5.

Gartner [8] predicted that around 26 billion devices would be connected to the network by 2020 and these would generate a huge amount of electronic waste and energy consumption. The need of the hour will be to reduce energy consumption and develop green communication techniques. This new trend is better known as Green IoT (G-IoT). Also, energy consumption is a severe issue in diverse heterogeneous IoT devices as it vigorously affects the budget and obtainability of the IoT network. Thus, energy depletion has developed as an essential concern and diverse algorithmic methods have been instigated to supplement hardware or dissimilar system-based approaches [22].

4.3 OVERVIEW OF GREEN IoT

The Internet of Things (IoT) encompasses the massive predictable evolution of the net in practice and of the nodes in prospect. Consequently, reduction in cost is required. Energy depletion has to be tackled in order to attain a Green IoT dependability and smart world employment. To deliver a sustainable smart world, the IoT

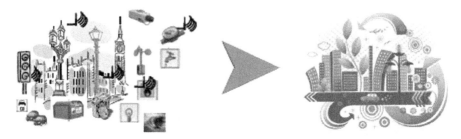

FIGURE 4.6   Green IoT

should demonstrate energy efficiency to diminish the greenhouse effects and carbon dioxide (CO_2) emissions from sensors, devices, applications, and services.

Green IoT [8] is defined as a set of processes espoused by the IoT for hardware or software procedures to attain energy efficiency through lessening of greenhouse effect in existing amenities and applications. Green IoT is spread through the association of facilitating skills, communication policies, and procedures and is depicted in Figure 4.6.

General ethics of Green ICT involves the following.

- Switch off the services that are not required.
- Disperse the required data only.
- Minimize span of information path by deploying energy efficient routing schemes.
- Minimize length of wireless data path by using energy efficient architectural strategy.
- Interchange handling of public services by utilizing data fusion.

4.3.1   GREEN IoT LIFECYCLE

The Green IoT lifecycle [4, 8] aims to have a minimal impact on the environment and has briefly described constituents as shown in Figure 4.7.

- **Green design**: planning energy competent arrangements for green IoT like audio constituents, PCs, and servers and cooling apparatus.
- **Green use**: diminishing power depletion of PCs and extra information schemes and expending them in an environmentally sound manner.
- **Green manufacturing**: creating automated constituents and laptops and new connected subsystems with negligible or no influence on the environment.
- **Green disposal**: renovating and refurbishing ancient PCs and reprocessing unwanted PCs and additional electronic tools.

4.3.2   TECHNOLOGIES FOR GREEN IoT

IoT is a new incipient knowledge that simplifies information communication amid manifold electronic devices deprived of social and PC intrusion. Green IoT [11] is a

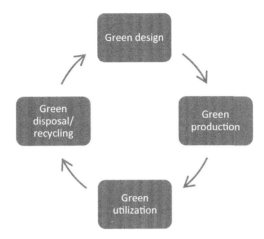

FIGURE 4.7 Lifecycle of Green IoT

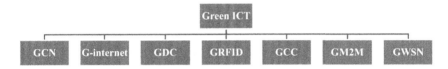

FIGURE 4.8 Categories of Green ICT

set of processes accepted by the IoT in the method of hardware or software efficiency procedures. This section primarily emphasizes the vital communication approaches and skills that contribute to Green IoT [10].

Owing to the mounting responsiveness to ecological concerns, Green IoT expertise creativities ought to be brought into the discussion. Greening IoT calls for skills that mark the IoT milieu in an amicable style by creating amenities and storage that permits the subscribers to collect, accumulate, access, and use numerous data. The technologies for Green IoT depend on Information and Communication Technology (ICT) expertise. The plans for sustainability of ICTs have focused on data center optimization through restructuring, leading to an increase in energy proficiency, diminishing CO_2 emissions and e-waste [13]. Consequently, greening ICT equipment plays an indispensable part in Green IoT and delivers numerous benefits to humanity. Greening ICT [14] is a combination of several enabling technologies as shown in Figure 4.8.

Green RFID technology (GRFID) Green RFID technology is a combination of radio frequency wireless communication technology and tag identification information. It is an automated data collection method which is able to plot actual world information easily into the graphic domain. It uses radio waves to reclaim, recognize, and collect information distantly. Components of RFID are RFID Tag (carrying and identification of data), Reader (read and write tag data and identifiers with a system), antenna and station (process the data).

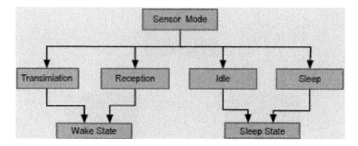

FIGURE 4.9   Sensor nodes for Green IoT

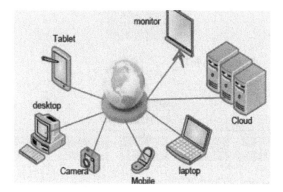

FIGURE 4.10   Green cloud computing

Green Wireless Sensor Network Technology (GWSN) WSN is the chief knowhow where a sensor is a blend of a huge amount of small, low-control, and low-cost automated devices. They transmit a signal continuously and have a sleeping mode when not in use for minimal power usage. The main aim of WSN is to augment the system life and contribute to enhance QoS parameters [27]. It is shown in Figure 4.9.

Green Cloud Computing (GCC) Its main role is to encourage the exploitation of ecological products which are simply reprocessed and salvaged. Its promising purpose is to diminish the usage of threatened resources, make the best use of energy consumption, and boost the recyclability of obsolete products and trash to minimize power requirements [27]. It is depicted in Figure 4.10.

Green Machine to Machine technology (GM2M) The main purpose is to reduce power consumption by employing numerous machines which can share information and collaborate without human intervention.

Green Data Center Technology (GDC) It reduces the power consumption of data centers, preserves quality of service requirements, minimizes energy, and has efficient bandwidth management.

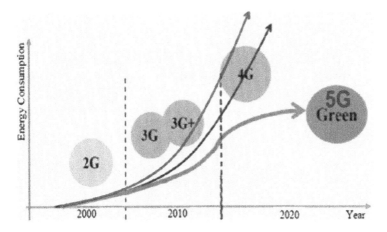

FIGURE 4.11   Developments of energy consumption for green communication

Green Communication and Networking (GCN) This is maintainable, energy responsive, energy-efficient, and environmentally aware expertise for low CO_2 emissions, and less coverage by radiation. 5G technology [16-21] permits great exposure to connectivity, lessening latency, saving energy, and preserve progressive data rates and system capability [27].

Green Internet Technologies (G-Internet) The internet notions and skills have contributed to progress of a smart and green grid, thus decreasing power depletion. It helps to simplify the act of planning transportation and routers.

4.3.3   APPLICATIONS

Numerous applications [4, 8] and services of Green IoT exist. Some of them are listed below in Figure 4.12.

4.3.4   ISSUES AND FUTURE INVESTIGATION DIRECTIONS

Though there have been remarkable investigation efforts to attain a green technology, Green IoT technology is still in in an early phase. There are numerous hindrances and [8] difficulties that must be handled. Below, we mention the main challenges:

- To attain acceptable performance, incorporation amid energy efficiency architectures is mandatory.
- There must be a minimum effect of applications on the environment.
- There should be consistency of Green IoT with energy consumption prototypes.
- Mutually the devices and procedures used to converse must be energy efficient with a reduced amount of power depletion.
- There should be a reduction in complexity of the Green IoT structure.

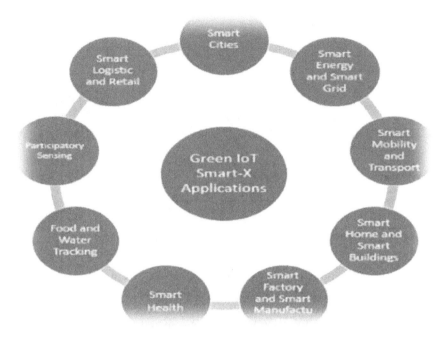

FIGURE 4.12   Green IoT applications

- Balance is needed between competent dynamic spectrum sensing and well-organized spectrum management.
- Efficient energy mechanisms for IoT such as wind, solar, and thermal make the IoT promising.
- Effective cloud management in terms of power consumption is needed.
- There should be efficient safety measures such as encoding and control directions.

There are many challenges in switching from IoT to Green IoT. These can be based on different parameters like software, hardware, routing algorithms, policy recycling, and so on. These should be designed in such a way as to save energy.

4.4   SMART CITY OVERVIEW

The foremost objective of smart cities [22] is to improve the outmoded facilities that are delivered to the inhabitants and also build new and more sophisticated ones. This visualization targets not only residents' affluence, but also the financial development and sustainability of the town. It is possible to attain this objective through the usage of skills and designs which seek to assimilate the several essentials of the town and aid them to interrelate in an operative way. In this section, we consider the main knowhow, characteristics, features, applications, and architecture that have to be realized for the expansion of smart cities to provide an integrated context and for easy understanding of smart cities. A smart city overview is shown in Figure 4.13.

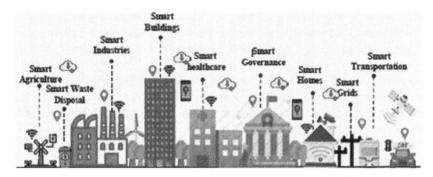

FIGURE 4.13 A smart city scenario

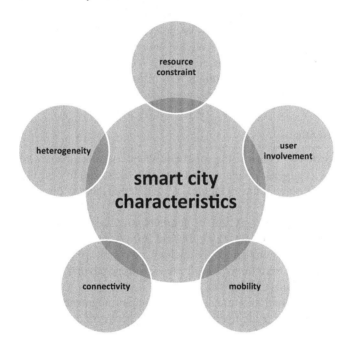

FIGURE 4.14 Features of smart cities

The city developments in a global scale is the main cause because it is expected that 60 percent of the world population will be living cities by 2025, with 55 percent in emerging countries by 2023. The main issue for building a smart city [30] is to reduce costs and resource consumption and the solution is Green IoT which aims to provide sustainability and low carbon emissions.

4.4.1 CHARACTERISTICS

The following are certain traits of smart cities [4] that ought to be deliberated and united before any different safety or privacy security process emerges. They are shown in Figure 4.14.

Heterogeneity: High heterogeneity refers to independent, distributed, or multi-user systems. There is no one-size-fits-all description of a smart city. IoT design differs according to the particular city. The absence of a collective safety agenda and service causes an issue.

Resource constraints: The IoT devices are resource inhibited, with restricted memory, battery capability, handling capabilities, and network constrained interfaces. Thus energy inefficient devices which are cheap, small, specific, with low random access memory are widely applied in smart cities.

Mobility: Mobility in smart cites is modified via the well-built communication system and refers to movement of data and goods from one place to another, wireless communication, and concurrent observation of the traffic flow.

Connectivity and scalability: Connectivity is the vital feature which permits any device to connect to the smart world for developing smart cities [24]. Scalability is a noticeable trait in smart city set-ups as they are speedily emerging from trivial to enormous, leading to an explosive evolution in both data and web traffic.

User involvement: Smart cities are mainly focused to serve the residents and improve the standard of smart solutions. For instance, preliminary consideration of security necessities will lead to notable results in terms of safety measures.

4.4.2 APPLICATIONS

The chief objective of constructing smart cities is to assist residents with their needs for energy, environment, commerce, and services. We exemplify the evolving intellectual uses [4] of smart cities in Figure 4.15 and define them in detail as follows.

1. Smart government: A smart administration using interconnected data, institutions, proceedings, and physical infrastructures based on information technology aids citizens to take decisions and make plans, thereby improving efficiency and transparency. For instance, online payment of bills or reporting of problems aids citizens [12, 13].
2. Smart transportation: This can aid the public by augmenting safety, swiftness, and consistency using intelligent transport networks, help in finding optimal route in terms of cost and speed, help in issuing passports, license recognition systems, car parking. [14] [15].
3. Smart environment: Smart milieu can contribute considerably in constructing a sustainable society [16]. New eco-friendly sensor nets might have the capability to forecast and identify imminent natural disasters [17].
4. Smart utilities: Smart services permit smart cities to decrease the overconsumption of assets like water and air and to increase monetary development and pay for ecological safety. Smart metering, as a real-world smart efficacy use, is extensively used in smart networks to observe the use of these assets [18]. Smart water meters [19] and smart light sensors [20] are employed to preserve assets and lessen damage.
5. Smart services: These profit people in several stages. For example, remote control of home appliances can allow positive, energy-saving situations, smart shopping, and entertainment and people are able to monitor health conditions via medical sensors and wearable devices.

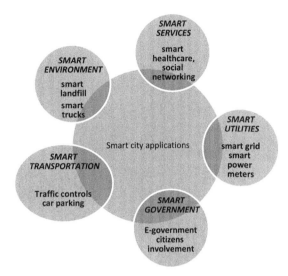

FIGURE 4.15 Uses of smart cities

4.5 FEATURES OF GREEN IOT SMART CITIES

To diminish the resources and energy consumed, the following are the main criteria for Green IoT reliability and smart world implementation which thereby reduce greenhouse effects and CO_2 emission of sensors, devices, services, and applications [22].

4.5.1 GREEN SMART HOME

This will facilitate different prospects for designers, builders, experts, and landowners. Novel sensor constructed home merchandises are manageable from the owner's smart sensory appliances such as smartphone or tablets – governing air conditioners, lighting, opening doors, and other domestic tasks.

The main requirements for green smart home [23] are actual interval energy observing tools to provide responses concerning energy usage and price, good power management to automatically switch on and off machines, restructured enclosed thermostats to restrict the carbon imprint, smart home lighting system making homes more eco-accommodating, smart water system like Cyber rain to profoundly adapt to any dimension.

Many service providers have launched residential applications controlled via a set-up box or the web to allow their clients to monitor their houses from anywhere via smart mobile phones. For instance:

- Apple Inc. has developed a Home Kit to allow people control their homes.
- Google has launched a home automation solution called Nest, which learns from daily habits and programs and schedules to provide security and an energy saving tool.

FIGURE 4.16   Greensmart home

- iRobot Corp developed a robotic based technology solution called Roomba, to clean floors of an entire house using integrated adaptive navigation and visual localization techniques.
- LG has introduced a natural language processing software termed HomeChat to connect people and communicate with their smart homes.

4.5.2   GREEN SMART OFFICE

Green IoT is a novel perception for planning smart offices which delivers these advantages:

- Improves business norms.
- increases returns on investment.
- Diminishes operating costs.
- Delivers improved space and resource supervision.
- Provides better efficiency.

Elucidations or approaches to decrease energy use are:

- Fit dimmer switch to lights.
- Artificial light can be tuned to the availability of natural light.
- Put in motion sensor lighting: switch off lights when nobody is in workplace.
- Connect programmable regulators: it modifies heat during business hours and avoids wasting energy in non-business hours.

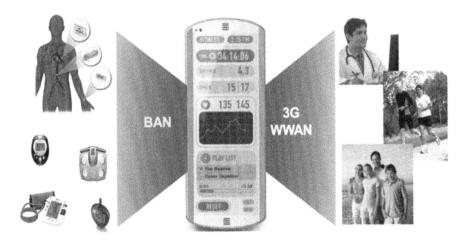

FIGURE 4.17 Green smart healthcare systems

4.5.3 GREEN SMART HEALTHCARE SYSTEMS

Healthcare can be the major recipient of the IoT revolution. Its chief goal is to promote disease prevention policy and improve health conditions. It comprises:

- Sensors to gather patient info.
- Internet access to convey information.
- Cloud computing to store and use data, and analyze information.
- Web and mobile solicitations to control data.

Case studies include Adidas Smart Run which has GPS to monitor heart rate of athletes. Softweb solutions include Pebble watch to manage emergency contacts, medicine alerts, and notifications for medicine, and other activities.

4.5.4 GREEN SMART FARMING

This encompasses mechanisms to augment the quantity and develop the superiority of crops. Smart farming embraces numerous ideas like use of water, insecticides, and dung, harvest processes, and examination of ecological effects. It is like a third green revolution with consolidated utilization of ICT arrangements like IoT instruments, GPS, big data, and mechanical autonomy.

4.5.5 GREEN SMART GRIDS

It is defined as a dispersed web-created scheme which provides improved governance of prevailing grid technologies and is a combination of traditional electric power grid and recent IT technologies. It combines novel resources with present operational

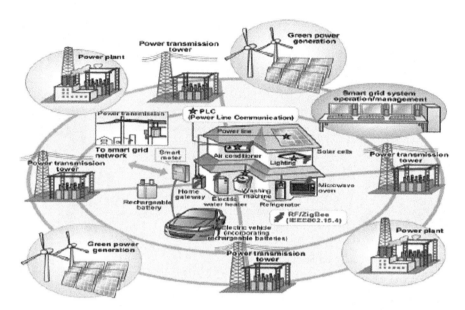

FIGURE 4.18   Green smart grid

schemes and helps to reduce resource utilization and energy consumption and exchange the power generated for new management options [24].

4.5.6   GREEN SMART TRANSPORT SYSTEM

For an intelligent transport system, infrastructure design must include smart and green technologies. The issues and challenges are related to tax charges, traveler baggage, lodging commercial carriers, by supporting the security arrangements of the transportation.

4.5.7   GREEN SMART WASTE MANAGEMENT SYSTEM

In a green smart waste management system, the waste to energy processes can be composted, recycled, or converted into energy, biofuel, or biochar through a lifecycle [25].

4.5.8   GREEN SMART ENVIRONMENT

The increasing quantity of automobiles, urbanization, and widespread contemporary practices have greatly increased air pollution in recent decades. Monitoring air pollution is a supremely tedious job. In progressive green smart cities, the deployment of IoT inventions can make observation of air pollution less complex [26].

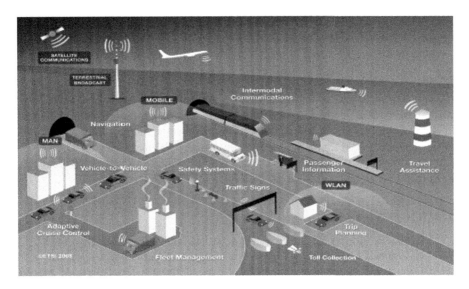

FIGURE 4.19 Green smart transport system

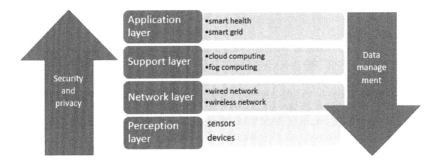

FIGURE 4.20 IoT based architecture for a smart city

4.6 IOT STRUCTURAL DESIGN FOR SMART CITIES

To keep improving smart cities, several architectures [27] have been suggested [28]. Still, there is no unvarying IoT design. A well-known and accepted four-layered architecture is briefly discussed [29] as shown in Figure 4.20.

Perception layer or the sensing layer is the lowest layer used for information collection from things prevailing in the physical layer and transferring to the network layer for execution.

Network layer is the vital layer, being influenced by nets, like the internet, WSNs, and communications networks. The duty of this layer is to assign the data stored by the perception layer and to connect smart things, network elements, and servers.

Support layer, maintains the necessities of differentiated applications via intellectual computing procedures.

Application layer is the topmost layer and delivers virtual and real-world amenities or tenders to clients constructed for their needs.

4.7 CONCLUSION AND FUTURE DIRECTIONS

IoT signifies an alteration in ICT that will permit smart cities to be developed everywhere. Green cloud computing is different from conservative computing and aims to include energy-efficient technology and improve waste and recycling procedures. In this chapter, IoT technology has been discussed.. The incentives for Green IoT, issues, and benefits are presented. The technologies involved in IoT and future investigation directions and issues are also reviewed.

BIBLIOGRAPHY

[1]   R. Ahmed, M. Asim, S. Z. Khan, and B. Singh, "Green IoT issues and challenges," *2nd International Conference on Advance Computing and Software Engineering"*, 2019.

[2]   J. Shuja *et al.*, "Greening emerging IT technologies: Techniques and practices," *Journal of Internet Services and Applications"*, Corpus ID: 20306034, 8(1), 2017.

[3]   I. Ganchev and Z. Ji, "A generic IoT architecture for smart cities," *ISSC 2014/CIICT*, 2014, pp. 196–199.

[4]   M. A. M. Albreem *et al.*, "Green Internet of Things (IoT): An overview," *4th Int, Conference on Smart Instrumentation, Measurement and Application*, November 2017, pp. 1–6.

[5]   S. Saxena and D. Saxena, "Green cloud computing architecture with efficient resource allocation system," *International Journal of Trend in Research and Development*, 3(6) 2016, pp. 248–251.

[6]   I. Chunsheng Zhu, V. C. M. Leung, and E. C.-H. Ngai, "Special section on challenges for smart worlds," *IEEE Access*, 3, 2015, pp. 2151–2162.

[7]   V. Subbarao , K. Srinivas , and R. S. Pavithr, "A survey on Internet of Things based smart, digital green and intelligent campus," *IEEE*, 2019, pp. 1–6.

[8]   T. H. Szymanski, "Security and privacy for a green Internet of Things," *Trust. Internet Things*, 19(5), 2017, pp. 34–41.

[9]   M. Z. F. Noshina Tariq, M. Asim, F. Al-Obeidat, M. Hammoudeh, and I. G. Thar Baker, "The security of big data in fog-enabled IoT applications including blockchain: A survey," *MDPI*, 19(8), 2019, 1424–8220.

[10] Rafeeq Ahmed, Bharat Singh, and M. Asim, "Green IoT—Issues and challenges," in *2nd International Conference on Advanced Computing and Software Engineering (ICACSE-2019)*, 2019, pp. 378–382.

[11] Saurabh Singh, In-Ho Ra, W. Meng, and G. H. Cho, "SH-BlockCC: A secure and efficient Internet of Things smart home architecture based on cloud computing and blockchain technology," *International Journal of Distributed Sensory Networks*, 15(4), 2019, pp. 18.

[12] C. Kyriazopoulou, "Smart city technologies and architectures," in *4th International Conference on Smart Cities and Green ICT Systems*, 2015, pp. 1–12.

[13] J. Shuja, R. W. Ahmad, and A. Gani, "Greening emerging IT technologies: Techniques and practices," *Journal of Internet Servers*, 8(9), 2017. https://doi.org/10.1186/s13174-017-0060-5

[14] L. T. Y. Chunsheng Zhu, L. Shu, V. C. M. Leung, S. Guo, and Y. Zhang, "Secure multimedia big data in trust-assisted sensor-cloud for smart city," *IEEE Communications Magazines*, 2017, 55 (12), pp. 24–30.

[15] R. Khatoun and S. Zeadally, "Cybersecurity and privacy solutions in smart cities," *Enabling Mobile Wireless Technology in Smart Cities*, 2017, 55(3), pp. 51–59.

[16] I. Lei Cui, G. Xie, Y. Qu, and A. Y. Y. Longxiang Gao, "Security and privacy in smart cities: Challenges and opportunities," *IEEE Access*, 6, 2018, pp. 46134–46145.

[17] K. M. H. M. L. Minh Dang, M. J. Piran, D. Han, "A survey on Internet of Things and cloud computing for healthcare," *MDPI*, 8(768), 2019.

[18] J. Dutta. and S. Roy, "IoT-fog-cloud based architecture for smart city: Prototype of a smart building," *7th International Conference on Cloud Computing, Data Science & Engineering-Confluence*, 2017, pp. 237–242.

[19] Y. Qian., D. Wu., W. Bao, P. Lorenz, "The Internet of Things for smart cities: Technologies and applications," *IEEE Network*, vol. 33, no. 2, 2019, pp. 1–5.

[20] J. R. Kuan Zhang, J. Ni, K. Yang, and X. Liang, "Security and privacy in smart city applications: Challenges and solutions," *7th International Conference on Cloud Computing, Data Science & Engineering-Confluence*, 55(1), , 2017, pp. 122–129.

[21] A. Alabdulatif, I. Khalil, H. Kumarage, and X. Yi, "Privacy-preserving anomaly detection in the cloud for quality assured decision-making in smart cities," *Journal of Parallel and Distributed Computing*, 127, 2017, pp. 209–223,.

[22] Kiran Kumar Vadlamudi, Ch. Ravi Kumar, "An enhanced methodology on Internet of Things with cloud in smart electrical systems," *International Journal of Recent Technology and Engineering*, 8(3), 2018, pp. 2295–2299.

[23] Sarder Fakhrul Abedin, Md. G. R. Alam, and R. Haw, "A system model for energy efficient green-IoT network," *ICOIN*, 2015, pp. 177–182.

[24] I. M. Marjani, F. Nasaruddin, A. Gani, I. A. T. Hashem, and A. I. Yaqoob, "Big IoT data analytics: Architecture, opportunities, and open research challenges," *IEEE Access*, 5, 2017, pp. 5247–5261.

[25] P. H. Zubair, A. Baig, P. Szewczyk, C. Valli, P. Rabadia, K. S. Maxim Chernyshev, M. Johnstone, P. Kerai, A. Ibrahim, and M. P. Naeem Syed, "Future challenges for smart cities: Cyber-security and digital forensics," *Digit. Investig.* 2017, pp. 1–11.

[26] D.-N. Le, B. Seth and S. Dalal, "A hybrid approach of secret sharing with fragmentation and encryption in cloud environment for securing outsourced medical database: A revolutionary approach", *Journal of Cyber Security Mobility*, 7(4), 2018, pp. 379–408.

[27] B. Montgomery, "Future shock by Bill Montgomery IoT benefits beyond traffic and lighting energy optimization," *IEEE Consumer Electronics Magazine*, 2015, pp. 98–100.

[28] S. H. Alsamhi, O. Ma, M. Samar Ansari, and Q. Meng, "Greening internet of things for smart everythings with a green-environment life: A survey and future prospects," *Telecommunication Systems*, 2018, pp.1–14.

[29] L. Atzori, A. Iera, and G. Morabito, "The internet of things: A survey", *Computer Networks*, 54, 2010, pp. 2787–2805.

[30] J. Gubbi, R. Buyya, S. Marusic, and M. Palaniswami, "Internet of Things (IoT): A vision, architectural elements, and future directions," *Future Generation Computer Systems*, 29, 2013, pp. 1645–1660.

[31] D. Popa, and M.-M. Codescu, "Reliabilty for a green Internet of Things", *Buletinul AGIR*, 2017, Corpus ID: 208317016, pp. 45–50.

[32] S. S. Prasad, and C. Kumar, "A green and reliable internet of things" *Communications and Network*, 5, 2013, pp. 44–48.

5 Green IoT and Big Data
Succeeding towards Building Smart Cities

Jyoti Nanwal, Puneet Garg, Preeti Sethi, and Ashutosh Dixit

CONTENTS

5.1 INTRODUCTION TO INTERNET OF THINGS

The term "Internet of Things" was first proposed by Kevin Ashton in a presentation in 1998 [1] where he mentioned that "The Internet of Things has the potential to change the world, just as the Internet did. Maybe even more so" [1]. IoT is a network formed by interconnection of not only communication devices but various other physical objects like cars, computers, home appliances, thereby transforming the world into a small village. IoT has been envisaged as covering a number of technologies and the research describes a number of technologies and research disciplines that enable global connectivity via physical objects. Within a short span of time, IoT has grown in multiple dimensions, which encompass various networks of applications, computers, devices, as well as physical and virtual objects, that are interconnected together using communication technologies such as wireless, wired and mobile networks, radio-frequency identification (RFID), GPS systems, and other evolving technologies. It is thus an integration of various technologies like radio-frequency identification, sensor networks, nanotechnologies, and biometrics. The plethora of applications being covered by IoT include intelligent transportation, e-health, smart home, wearables, smart homes, connected cars, smart retail, to name a few. With the advancement of technology, there is a rapid need for solutions to connect billions of devices in hospitals, homes, cars, factories, and many other places. Also there is a need to collect, store, and then analyze the data obtained from the devices. So there is a need to build IoT solutions for a wide range of devices.

5.1.1 Green Internet of Things

Though IoT concepts have managed to interconnect devices, this is achieved at the cost of high power consumption. Due to ubiquitous IoT, high power consumption is an issue which of upmost concern to the researchers. There is an urgent need to develop an energy efficient procedure for connecting devices, not only reducing power consumption but also reducing the greenhouse effects of IoT and minimizing emission of CO_2 so that there will be very small impact on the environment and the environment will be safe enough to live in. This modified energy efficient IoT is referred to as Green IoT (Green IoT).

Green IoT entails user friendly, energy efficient, advanced, smart, and sustainable devices. It aims to

- Introduce green computational units, network-based architectures and communication protocols with comparatively low energy utilization and there should be maximum utilization of bandwidth.
- Use techniques to diminish carbon emissions and pollution. There should be consumption and disposal of green units for efficient utilization of energy.

According to [8], Green IoT is predicted to introduce significant changes in our daily life and help realize the vision of "green ambient intelligence." In a short time we will be communicating via 5G and acting "intelligently" with help of a huge number of sensors, devices, and these will manage user's tasks by providing green support [8]. Green IoT can be achieved if we follow the greening i.e. energy saving/ minimizing approach from the design phase itself. Devices should go into sleep mode when not being used, best routing algorithm should be used for data exchange so that cost and energy consumption can be minimized, use energy that is freely available (renewable green power source).

Green IoT involves several technologies such as green RFID tags, green sensing network, and green cloud computing network.

Green IoT aims to preserve natural resources by minimizing the technology impact on the environment and human health [9]. To achieve this objective, Green IoT ensures that the complete cycle ranging from design, production, utilization, and recycling is energy efficient [10].

1. *Green design*: Designing energy efficient components, computers, and servers and cooling equipment.
2. *Green manufacturing:* Producing electronic components and computers and other associated subsystems with minimal or no impact on the environment.

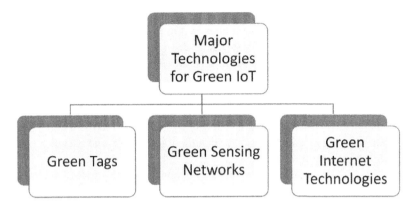

FIGURE 5.1   Key technologies for Green IoT

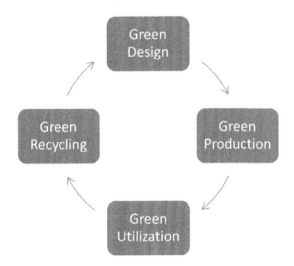

FIGURE 5.2   Lifecycle of Green IoT

3. *Green use*: Minimizing power consumption of computers and other information systems as well as using them in an environmentally sound manner.
4. *Green disposal:* Refurbishing and reusing old computers and recycling unwanted computers and other electronic equipment.

5.1.2 APPLICATION OF GREEN IOT

Smart Cities

The Green IoT initiative is the consequence of endeavors to improve the standard of living by means of ecological insurance and manageability utilizing innovation. One of the biggest aims is to equip cities with technology but for them to be sustainable places to live. The network of systems with devices and infrastructure in all metro cities across the world enables stakeholders to diminish carbon emissions, water utilization, increase security, productivity, and human prosperity [2]. The following are the features which contribute to make smart cities:

- **Smart parking**: Smart parking is an attempt to reduce traffic congestion faced during a conventional parking process. It is the real-time monitoring of vehicles to identify and reserve available parking spaces nearest to them in the city.
- **Smart lighting:** This refers to including intelligent and weather adaptive lighting in street lights.
- **Safe city: This** is the use of digital video monitoring, fire control management, public announcement systems, in a given city.
- **Intelligent transportation system:** This involves building smart roads and intelligent highways with warning messages and diversions in the face of weather issues and unexpected events like accidents or traffic jams.
- **Waste management:** With the ever-increasing waste products which are dumped every day, there is an urgent need to synthesize the collected waste optimally. The Green IoT based waste management system aggregates the data from multiple sensors and thereafter transmits them to the cloud with the goal making investigation, anticipating, and enhancement activities more efficient.

Smart Grid

Smart grid is a communication network developed to collect and analyze data from various parts of a power grid so that the power supply and demand can be predicted [14]. It has been developed as a means to deal with high carbon emissions, high greenhouse gas, and issues of safety, security, related to conventional power grid. Smart grid uses various renewable energies and storage systems like plug-in electric cars with bi-directional chargers, batteries that are connected to distributed systems.

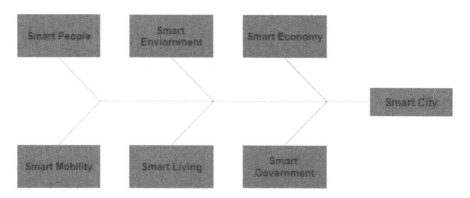

FIGURE 5.3 The smart city

Smart Transport

Intelligent transportation system (ITS) is a successful transportation and mobility system utilized in smart cities. It exploits innovations, for example, the Internet of Things (IoT) and big data analytics to oversee traffic and mobility, upgrade the transport framework, and give improved interfaces to transport administrations. Benefits of an ITS incorporate making connecting transport systems with open correspondence among devices and vehicles, effectively overseeing traffic, helping transport to keep on time, and furthermore guarantees citizens have access to real-time information about traffic and transportation conditions [37]. ITS promotes smoothing of traffic flows on the road network. Increased use of green transportation choices can also limit the need for expensive investment in infrastructure and minimize unpopular traffic regulations.

Smart Manufacturing and Smart Factories

The advanced manufacturing industry is putting resources into new technologies, for example, the Internet of Things (IoT), big data analytics, distributed computing, and cybersecurity to adapt to system complexity, increase information visibility, improve production performance, and gain recognition in the worldwide market. These advances are quickly empowering another age of smart manufacturing, i.e., a cyber-physical system firmly integrating manufacturing undertakings in the physical world with virtual ventures in cyberspace. To a large extent, understanding the full potential of cyber-physical systems depends on the development of new methodologies in the Internet of Manufacturing Things (IoMT) for data-enabled engineering innovations. The Industrial Internet of Things (IIoT), smart factory, and smart manufacturing are the applications of the IoT that focus on cheap, smart, and small-size interconnected factory devices [16].

Smart Health

Online healthcare applications empower the sharing of patient-specific records and well-being data whether a patient is at home or at the emergency clinic. Healthcare sensor devices associated with patients' screen their BP, temperature, heart rate, and so forth, and track their movement and conduct so as to send data to the medical clinic systems as a major aspect of the healthcare process [11].

An attacker could exploit frail validation instruments to increase unapproved access to systems such as embedded web servers in the clinic building. This could permit the attacker not just to invade a portion of the emergency clinic system and gain unauthorized access to sensitive information, but additionally empower them to channel their way through to the sensors so as to target patients [11]. IoT medical sensors are designed to gather data ceaselessly on a patient's status and medical clinics use that data to help their patients as a feature of their healthcare monitoring and for data examination purposes. Some green IoT application in building smart health are:

- **Physical activity monitoring for the elderly**: For measuring motion and vital signs of a person we can use body sensors and can record activity data.

- **Patient's surveillance:** Inside the hospitals or in the old people's care homes we can monitor patient's activities to avoid dangerous situations.
- **Chronic disease management:** Patient monitoring systems with comprehensive patient statistics could be available for remote residential monitoring of patients with chronic conditions such as pulmonary and heart disease and diabetes.
- **Sleep control:** We can monitor mattress stresses and small motions while sleeping by using wireless sensors placed across the bed, just as we can monitor breathing patterns and heart rate. Through a smartphone application we can also monitor motions caused by tossing and turning during sleep hours.

Some other applications for green IoT are food and water tracking, participatory sensing, logistics, and retail.

5.2 INTRODUCTION TO BIG DATA

Big data is a term used for voluminous data used for decision-making after some pre-defined analysis methods. To understand the term big data, we may consider that it to be composed of five Vs.

All five Vs can be understood one by one. The first V is **voluminous**; it means the data size is enormous. Data are considered Big Data if and only if they are very huge in size. Second V stands for **velocity,** which means the rate at which data are being generated –they should be massive and continuous in nature. For example, more than 3.5 billion searches per day are made on the internet. **Variety** refers to the nature of data coming from different sources that could be structured, semi-structured, and unstructured. **Veracity** refers to data which are inconsistent and uncertain – sometimes data become messy or noisy and come from different sources so controlling those data becomes difficult. If half the data can be analyzed, this will not provide useful information (i.e. the information obtained can be true or false) and if we analyze too much data then the result we get will give us uncertain results. Fifth V stands for **value** – that is the most important characteristic of big data as data are of no use until we find some patterns from them and convert them into valuable information.

At every mouse click, lots of information can be obtained. By analyzing that information on customers' buying patterns they can then be recommended the same products in the future. With the help of comments and tweets generated by social media such as Facebook and Twitter we can analyze and see reviews on a particular product or on schemes. We can also use the analysis of images, videos, and audios like facial recognition in real-world applications.

5.2.1 APPLICATIONS OF BIG DATA

Big Data offers numerous applications in the real world. Some of them are listed below.

Health care

Big data have brought a drastic change in the health sector. Personalized health care services to individual patients have become possible due to predictive analytics by medical professionals. Apart from a variety of wearable health monitoring devices, telemedicine applications are also supported by big data. Traditionally, there was a limited ability to analyze big data so it was not easy to extract useful information from the previously obtained data. But now as big data are updated with the latest techniques, we can store huge amount of data using HDFS, do analysis of that collected data, and find some interesting patterns. Also, digital health records can reduce the need for paperwork. This record can be modified by a doctor if needed and includes the patient's medical history, demographics, all laboratory test information, which may be beneficial to doctor as well as for patient for further treatment and diagnosis.

Banking and Securities Industry

Big data are used for fraud detection in the banking sector. Big data tools can efficiently detect faulty alteration and misuse of credit card/debit cards. We can monitor financial market activities using big data. To capture illegal activities in the financial market, one can use network analytics and natural language processing. We can analyze large-scale patterns across several transactions and can detect anomalous behavior from individuals which can help in fraud detection.

Communications, Media, and Entertainment

A general website contains a high amount of information submitted by its customers. Based on that information, an organization creates a customer profile. This profile is used to provide relevant content to a targeted audience, recommend content, and also measure content performance. Likewise data are collected from millions of users and then processed to find some useful patterns to recommend videos to individual users. For example, Amazon Prime, YouTube are mainly based on big data.

Education

Education also is not limited by physical space nowadays. By using big data, various online courses are also available, helping students understand relevant topics. Many reputed institutes now focus on courses offered using big data. Huge amounts of data about the student are available on the university portal; they can analyze that available data and can find patterns to predict the performance of a student and help them maximize this. Also by analyzing the browsing histories and time spent on a particular site by students, one can find their topics of interest and what bores them.

Manufacturing and Marketing

Big data help in understanding customers' behavior around a particular product, allowing companies to plan and improve their product quality, so that the product can survive in the market. Also, we can make transparent infrastructure to predict incompetence and uncertainties which could be hazardous in the future. With the advancement of technologies in some places, a customer's facial expressions are also analyzed to know the reaction to the product, what makes them to buy that product and what makes them reject it. Capturing facial expression means capturing their feeling towards a certain product. So big data play a very important role in manufacturing to understand marketing strategies.

Agriculture

We always treat agriculture as an intuitive space where skills and techniques of farming pass from generation to generation, but it is also a fact that the population is increasing day by day, the climate is changing, and viable farmland is depleting, therefore, it is very important to step up our crop production. We need technologies which can yield more and better-quality crops in minimum space and this can only be possible by analyzing each factor affecting crop production at every second corresponding to changing climate. There is a need to take immediate action to yield more crops, so that the population which is growing at a fast pace can have something to eat.

We condense the uses of IoT procedures in agriculture into four categories: controlled environment planting, open-field planting, livestock breeding, and aquaculture and aquaponics [5]. The attention on putting agribusiness IoT systems into practice is proposed to be extended from the development cycle to the agri-item's life cycle. In the interim, a vital concern is the usage of agriculture IoT systems. The development of green IoT systems in the entire life cycle of agri-items will greatly affect farmers' interest in IoT methods.

There are some other applications, such as in smartphones there are facial recognition system technologies and there are certain applications by which you can take a picture and store some useful information about that person, so when your memory fails you, you can recognize that person with the help of that application. Nowadays travel agencies are also using data to better understand the customer's choices, suggest to them luxury hotels within their budget, provide them with transfers, and also the best deals or packages. Big data have become a perfect guide for a traveler. Telecom companies also use big data to analyze network traffic, thus ensuring hassle-free network connectivity. The data also help with smooth connectivity.

Big data are not limited to the mentioned applications only. Big data and artificial intelligence combine, offering driverless cars. With the help of big data, we can analyze the driving pattern of a driver and can make driving decisions based on that pattern. Therefore, there is no limit to the application of big data. Patient prediction for improved staffing is also a good example of using big data to avoid unnecessary labor costs.

5.2.2 BIG DATA TECHNOLOGIES AND TERMINOLOGIES

Hadoop Ecosystem

Hadoop is an open-source software framework for data. It stores data from a massive dataset on a cluster of low commodity machines in a distributed fashion. Hadoop ecosystem is required because the existing Relational Database Management System is not capable of storing and managing these huge datasets. If RDBMS were used, the cost of storing such huge amounts of data would be very high. Also in the real world, data can be available in any format, structured or unstructured or semi-structured, but the traditional system only stores structured data and is not capable of storing data in an unstructured or semi-structured format. another reason is to adopt Hadoop is that the data are increasing at rapid pace, The traditional RDBMS approach cannot handle this kind of voluminous data. With the help of the Hadoop system, sensitive data can be processed in real time in an efficient manner.

Cascading

This is an open-source data processing API for Java developers. Data processing workflow can be created and executed on Hadoop. It is the collection of applications. Cascading is for hiding the underlying complexity of map-reduce jobs. Cascading is a robust and data-oriented application in Java. It eliminates lock-in, creates higher-order DSL languages in other JVM based languages. Cascading offers portability and there is no cost to rewriting code if we move existing applications in cascading. We can simply change a few lines of code and we can easily port the cascading application.

Scribe

Scribe was developed by Facebook as an open source. In real-time scribe is used for aggregating long streams coming from different servers. It was scalable, robust of the network or any specific machine. It is nothing but a collection of processes that are running on parallel distributed machines and listening to a specified port from where the data are being inserted. Scribe works on files like write to a file or network and we can send data to another scribe server. We divide messages into different categories and then send them based on that category. Facebook was having some scaling challenges and scribe was developed to handle them. The power of scribe is to handle billions of messages per day.

Apache HBase

HBase is the column-oriented, distributed, non-relational database and scalable NoSQL database, it can be used when we need random real-time read/write access to data in the Hadoop distributed file system. In HBase, data are stored in the form of rows and columns and the table can extend to n number of rows and m number of columns where n, m are billions and millions, respectively. The intersecting point of row and column is called a cell and in each cell there is a time stamp identifier which is known as version. The table is row-wise sorted. There are multiple column families inside the table, there can be number columns inside a column family, and finally, column is the collection of key-value pairs. Apache HBase has three different parts namely HMaster, region server, and the zookeeper. For assigning regions to region server we use the HMaster process which also manages Hadoop clusters. By using HMaster we can create, modify, and delete tables of the database. Requests coming from the client go to region server where they are read, written, and being modified. The storage of the region server is divided into three parts—block cache, mem cache, and HFile. In reading cache read data are stored and when it is full recently used data are removed. Write cache stores the new data, that are not present in the disk. There will be a different write cache or mem store for each column family. The actual data are stored in the HFile on a disk. Zookeeper enables reliable distributed coordination; it is also an open-source server. It maintains configuration information and distributed synchronization.

Mango DB

It is an open-source, cross-platform, document-oriented NoSQL database which is written in C++ and created by founders of Double-click. It mainly stores structured types of data. It is adopted by MTV Networks, New York Times, etc. It is high in performance, high in availability, and easily scalable. It works on the concept of collection and documentation.

Spark

Spark is an open-source, distributed, general-purpose cluster computing framework for big data processing within Hadoop, It is a part of Hadoop but has a very important role. Therefore, it deserves a category of its own. It is 100 times faster than the Hadoop engine.

R

R is a programming language developed by Ross Ihaka and Robert Gentleman in 1993. Machine learning algorithm, linear regression, time series, and statistical inference are included in the R language. It is also openly available.

Data Lakes

Data lakes are nothing but the vast data repositories which make it easier to access the huge amount of data stores of many enterprises. Data are gathered from numerous heterogeneous sources and stored in their natural state.

NoSQL Database

Before NoSQL, RDBMS was used to store structured data defined in row and column. Developers and administrators of the database make queries manipulate and manage the data but as data are growing exponentially, RDBMS is failing to process these. NoSQL is a specialized database to store unstructured data. It does not provide the same level of consistency but provides very fast performance.

5.3   ROLE OF GREEN IOT AND BIG DATA FOR BUILDING SMART CITIES

As explained in the previous sections, Green IoT works in an ecofriendly manner to connect devices and enable them to share information. Big data add to the power to Green IoT by synthesizing the extracted information and providing valuable insight into the data being sensed. This analysis can contribute to meeting the challenges associated with building smart, sustainable, and greener cities, thereby increasing the quality of life and making this technology-equipped world a better place to live in.

The data collected by Green IoT enabled sensors are gathered from numerous sources and are too voluminous to be processed by traditional SQL-queried relational database management systems (RDBMSs). A platform to treat this huge amount of data is needed, so Hadoop comes into the picture. This approach is known as Fog

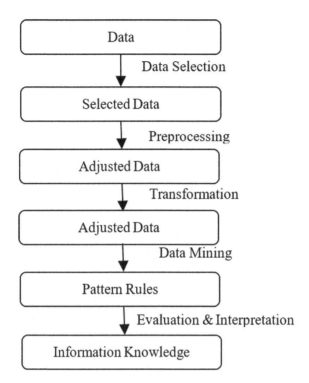

FIGURE 5.4 Extraction of knowledge

computing [22] and has considerably increased information saving, processing, and examining in the real-time environment.

The role of Green IoT and big data analytics in designing the smart and green city is explained as follows.

5.3.1 SMART METERS

Technological advancements have encouraged the deployment of smart meters that empower online estimating of consumption of different assets, for example, electricity, heat, gas, and water. Right now, smart meters create data essential for successful resource management. As such, big data analytics applied to data assembled from smart gadgets, sensors, and smart meters, empower consumption prediction, reduce blackouts, and help with resource monitoring. Green IoT components together with a cloud/fog/edge-based mechanized administration system hold the possibility to improve energy efficiency and water consumption. Based on big data analytics and visualizations, satisfactory predictions and forestalling actions can be taken along with other decisions in sustainable resource management [26]. In this way, the investigation of data produced by every electrical device and other smart devices utilized in the vision of smart and green buildings empowers the improvement of knowledge about the utilization of energy and different assets and prompts upgraded monitoring, controlling, and warning actions. For instance, smart grids in the future will be composed of micro-power system networks, connected to one another through the cloud, and able to monitor, run, or disconnect themselves and heal based on the data gathered by smart metering gadgets [27, 28]. In addition, utilizing creative building and automation systems, the consumption of energy in the building area can be diminished by 30–80 percent [29].

5.3.2 SMART TRASH MANAGEMENT SYSTEM

The data from multiple sensors and information systems, with useful information about container filling level, traffic congestion, and garbage truck location, are being collected and used for waste management system based on Green IoT. Then we transfer this collected data to the cloud so that various waste management companies can take those data, can perform analysis, and make predictions [30, 31]. The container can convey information on the waste level or ongoing collection and based on those data be set apart for collection. The garbage truck then only gathers full or overdue containers which simultaneously prompts the optimization of garbage truck routes. Right now, Green IoT fundamentally contributes to progressively improved, sustainable and cost-efficient waste administration.

5.3.3 SMART CITY SURVEILLANCE SYSTEMS

To fight crime and maintain public security there comes the role of city surveillance [32]. To make our city secure and sustainable we should use smart sensors, cameras, applications based on location and infrastructure, and we can improve our transport sector too. With the help of Green IoT we can also successfully control and optimize traffic. We can obtain at high speed a large amount of information from the roads, lights, connection of vehicles, and control systems. To make our traffic, driving, and parking more efficient, we need to gather and analyze real-time data [33, 34]. For instance, the Green IoT solutions based on data gathered and processed can prescribe the best route and ideal driving velocity to avoid congestion, or to assist drivers with getting the nearest accessible parking place in a busy zone [2]. The residents by means of their cell phones can generally have real-time information on public transport and its accessibility. Additionally, road lighting can be balanced by the proximity of individuals in a specific area, contributing to ensuring safety and minimized energy consumption simultaneously [35].

5.3.4 SMART EDUCATION

Education is one of the major domains which has undergone substantial change owing to the use of IoT enabled objects. The use of smart school labs, smart classrooms, smart transportation, smart monitoring systems are some of the most fascinating features which have modified school life – from the toddler to a young graduate [33].

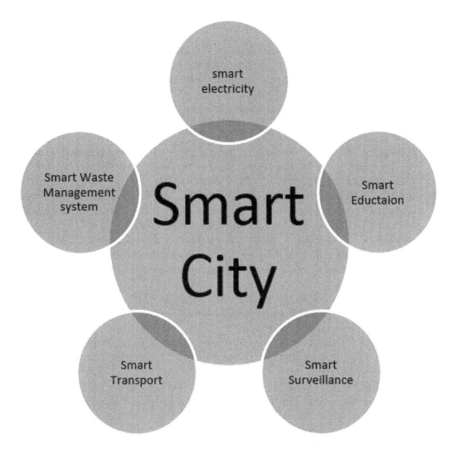

FIGURE 5.5   Glimpse of smart city

We have considered only a few of the numerous benefits of Green IoT with big data analytics towards smart city vision. Green IoT and big data are interrelated terms. To make intelligent decisions it is very important to study past data to derive valuable insights for enhancing the quality of life due to the use of enhanced smart city infrastructure and services.

5.4   SUMMARY

Detailed study presented in this chapter reveals that big data and Green IoT can go hand in hand and contribute towards making cities green and sustainable, rightly termed as smart cities. The chapter also reveals that the enormous amount of data being extracted can be used for analyzing and synthesizing. The integration of these two upcoming technologies has the power to revolutionize this technology-equipped world.

BIBLIOGRAPHY

[1]   Ashton, K. (2009). That "internet of things" thing. *RFID Journal*, 22(7), 97–114.

[2]   Maksimovic, M. (2017). The role of green internet of things (Green IoT) and Big Data in making cities smarter, safer and more sustainable. *International Journal of Computing and Digital Systems*, 6(04), 175–184.

[3]   Shoae, A. (2016). Green IoT solutions, smart city-smart environment-smart industry-smart metering solutions. www.linkedin.com/pulse/green-iot-solutionssmart-city-environment-industry-ali-shoaee

[4]   Gapchup, A., Wani, A., Wadghule, A., and Jadhav, S. (2017). Emerging trends of green IoT for smart world. *International Journal of Innovative Research in Computer and Communication Engineering*, 5(2), 2139–2148.

[5]   Ruan, J., Wang, Y., Chan, F. T. S., Hu, X., Zhao, M., Zhu, F., ... and Lin, F. (2019). A life cycle framework of green IoT-based agriculture and its finance, operation, and management issues. *IEEE Communications Magazine*, 57(3), 90–96.

[6]   Albreem, M. A., El-Saleh, A. A., Isa, M., Salah, W., Jusoh, M., Azizan, M. M., and Ali, A. (2017). Green internet of things (IoT): An overview. In *2017 IEEE 4th International Conference on Smart Instrumentation, Measurement and Application (ICSIMA)*, pp. 1–6. IEEE Malaysia.

[7]   Ahmad, R., Asim, M. A., Khan, S. Z., and Singh, B. (2019). Green IoT: Issues and challenges. In *Proceedings of 2nd International Conference on Advanced Computing and Software Engineering (ICACSE)*.

[8]   Al-Turjman, F., Kamal, A., Husain Rehmani, M., Radwan, A., and Khan Pathan, A. S. (2019). The Green Internet of Things (Green IoT). *Wireless Communications and Mobile Computing*.

[9]   Alsamhi, S., Ma, O., Ansari, M. S., and Meng, Q. (2018). Greening internet of things for smart everythings with a green-environment life: A survey and future prospects. arXiv, *arXiv preprint arXiv:1805.00844*.

[10]   Nandyala, C. S., and Kim, H. K. (2016). Green IoT agriculture and healthcare application (GAHA). *International Journal of Smart Home*, 10(4), 289–300.

[11]   Abouzakhar, N. S., Jones, A., and Angelopoulou, O. (2017). Internet of things security: A review of risks and threats to healthcare sector. In *2017 IEEE International Conference on Internet of Things (iThings) and IEEE Green Computing and Communications (GreenCom) and IEEE Cyber, Physical and Social Computing (CPSCom) and IEEE Smart Data (SmartData)*, pp. 373–378. IEEE, Exeter, UK.

[12]   Ghasempour, A. (2019). Internet of things in smart grid: Architecture, applications, services, key technologies, and challenges. *Inventions*, 4(1), 22.

[13]   Ghasempour, A. (2016). Optimized advanced metering infrastructure architecture of smart grid based on total cost, energy, and delay. In *2016 IEEE Power and Energy Society Innovative Smart Grid Technologies Conference (ISGT)*, pp. 1–6. IEEE, Minneapolis, MN.

[14]   Ghasempour, A. (2016). Optimum number of aggregators based on power consumption, cost, and network lifetime in advanced metering infrastructure architecture for Smart Grid Internet of Things. In *2016 13th IEEE Annual Consumer Communications and Networking Conference (CCNC)*, pp. 295–296. IEEE, Las Vegas.

[15]   NIST releases final version of smart grid framework. www.nist.gov/smartgrid/upload/NIST-SP-1108r3.pdf (accessed 16 January 2019).

[16]   Andersen, A., Karlsen, R., and Yu, W. (2018). Green transportation choices with IoT and smart nudging. In *Handbook of Smart Cities,* pp. 331–354. Cham: Springer.

[17]   Forrest, C. (2015). Ten examples of IoT and big data working well together. www.zdnet.com/article/ten-examples-of-iot-and-big-data-working-well-together/

[18] Hashem, I. A. T., Chang, V., Anuar, N. B., Adewole, K., Yaqoob, I., Gani, A., … and Chiroma, H. (2016). The role of big data in smart city. *International Journal of Information Management*, 36(5), 748–758.

[19] McLellan, C. (2015). The internet of things and big data: Unlocking the power. *ZDNet*. www.zdnet.com/article/the-internet-of-things-and-big-data-unlocking-the-power/

[20] Oracle (2013). Big Data Analytics – Advanced Analytics in Oracle Database. www. oracle.com/technetwork/database/options/advanced-analytics/bigdataanalyticswpoaa-1930891.pdf

[21] Chen, N., Chen, Y., Ye, X., Ling, H., Song, S., and Huang, C. T. (2017). Smart city surveillance in fog computing. In *Advances in Mobile Cloud Computing and Big Data in the 5G Era*, pp. 203–226. Cham: Springer.

[22] Cisco (2015). Fog computing and the Internet of Things: Extend the cloud to where the things are. www.cisco.com/c/dam/en_us/ solutions/trends/iot/docs/computing-overview.pdf

[23] Vujović, V., and Maksimović, M. (2015). Data acquisition and analysis in educational research based on Internet of Things. In *UDC 681.518 (04) INTERACTIVE S< STEMS: Problems of Human-Computer Interaction: Collection of Scientific Papers,* p. 57. Ulyanovsk: USTU.

[24] Foote, K. D. (2016). Techniques and algorithms in data science for big data. DataVersity. www.dataversity.net/techniques-and-algorithms-in-data-science-for-big-data/

[25] Ali, A., Qadir, J., Rasool, R., Sathiaseelan, A., Zwitter, A., and Crowcroft, J. (2016). Big data for development: applications and techniques. *Big Data Analytics*, 1(1), 2.

[26] ITU (2010). *ICT as an Enabler for Smart Water Management*, ITU-T Technology Watch Report. www.itu.int/dms_pub/itu-t/oth/23/01/T23010000100003PDFE.pdf

[27] GeSi (2015). *#SMARTer2030-ICT Solutions for 21st Century Challenges* Brussels: Global e Sustainability Initiative (GeSI). http://smarter2030.gesi.org/downloads/Full_report.pdf

[28] Vijayapriya, T., and Kothari, D. P. (2011). Smart grid: An overview. *Smart Grid and Renewable Energy*, 2(4), 305–311.

[29] Mohanty, S. P., Choppali, U., and Kougianos, E. (2016). Everything you wanted to know about smart cities: The internet of things is the backbone. *IEEE Consumer Electronics Magazine*, 5(3), 60–70.

[30] Medvedev, A., Fedchenkov, P., Zaslavsky, A., Anagnostopoulos, T., and Khoruzhnikov, S. (2015). Waste management as an IoT-enabled service in smart cities. In *Internet of Things, Smart Spaces, and Next Generation Networks and Systems,* pp. 104–115. Cham: Springer.

[31] Tracy, P. (2016). How industrial IoT is revolutionizing waste management. *Enterprise IoT Insights.* http://industrialiot5g.com/20160728/internet-of-things/waste-management-industrial-iot-tag31-tag99

[32] Talari, S., Shafie-Khah, M., Siano, P., Loia, V., Tommasetti, A., and Catalão, J. P. (2017). A review of smart cities based on the internet of things concept. *Energies*, 10(4), 421. doi:10.3390/en10040421

[33] Biswas, S. P., Roy, P., Mukherjee, A., and Dey, N. (2015). Intelligent Traffic Monitoring System. In *International Conference on Computer and Communication Technologies – IC3T*. Hyderabad, India.

[34] Solanki, V. K., Katiyar, S., BhashkarSemwal, V., Dewan, P., Venkatasen, M., and Dey, N. (2016). Advanced automated module for smart and secure city. *Procedia Computer Science*, 78(C), 367–374.

[35] KDnuggets (2015). How big data helps build smart cities. www.kdnuggets.com/2015/10/big-data-smart-cities.html

[36] Sierra Wireless (n.d.) Make cities safer and more efficient with smart city technology."
www.sierrawireless.com/applications/smart-cities/

[37] Gupta, S., Sethi, P., Juneja, D., and Chauhan, N. (2014). Software agent based forest fire
detection (SAFFD): A novel approach. *International Journal of Emerging Technologies
in Computational and Applied Science (IJECTAS)*, 8, 476–480.

6 Green Internet of Things (G-IoT) Application
AI Based Smart Farming

Shakti Arora, Vijay Anant Athavale, and Abhay Agarwal

CONTENTS

6.1 INTRODUCTION

"Smart farming" is an emerging concept that refers to managing farms using technologies like IoT, robotics, drones and AI to increase the quantity and quality of products while optimizing the human labor required by production [1]. Among the technologies available for present-day farmers are the following.

- Sensing technologies, including soil scanning, water, light, humidity, temperature management;
- Software applications,specialized software solutions that target specific farm types;
- Communication technologies, such as cellular communication;

- Positioning technologies, including GPS;
- Hardware and software systems that enable IoT-based solutions, robotics, and automation; and
- Data analytics that underlies the decision-making and prediction processes.

6.2 TARGETED AREAS AND OBJECTIVES

Smart farming applications do not just target large, conventional farming exploitations, but could also be new levers to boost other common or growing trends in agriculture, such as family farming (small or complex spaces, specific cultures and/or cattle, preservation of high quality or particular varieties), organic farming, and enhance a very respected and transparent farming in accordance with European consumer, society, and market preferences. Smart farming can also provide great benefits in terms of environmental issues, for example, through more efficient use of water, or optimization of treatments and inputs.

These innovation when installed on farms will certainly bear some sweet fruit after the successful execution and application of this technology. Our innovation can be applied to any farm irrespective of the crops and seasons. It will be cost-effective so that a farmer belonging to any particular tier can afford it easily and will have an easy user interface to make it more accessible thus multiplying the outputs.

Various projects and applications are deployed in agricultural fields leading to efficient management and control of various activities. In 2000 approximately 525 million farms were connected to the IoT, in 2016 540 million, in 2035 approximately 780 million, and by 2035 approximately 2 billion farms are likely to be connected [1, 2]

Introducing the new farming tech and fitting with existing techniques required on-the-ground knowledge of agriculture and crops. Digitalization of the farms meant taking on the various problems faced such as the water level, avoiding farm fires, quality checking of the crops by the new digital methods, and optimizing them in more reliable and efficient ways.

Farming is being practiced in India by the same old medieval techniques, with poor result. The agro-industry on the ground for the farmers is really decreasing, creating more troubles, so a change is needed. A project developed for the farming industry will help to improve things.

1. Target Industry: Our initiative targets all the people who are directly related to the agriculture industry or the farmers who are struggling with their farming business and looking for quality solutions.
2. Cost-Effective: Farmers are generally not well paid and hence their financial condition is not well suited to affording and adapting the technology. So our main focus is on cost-effective measures so that IoT based agriculture is not a noose round their necks.
3. User Friendly: The interface and handling of IoT devices are very user-friendly. Even a person without experience of using such similar devices may find it really smooth to use.

4. Terrain Friendly: Our innovation is terrain friendly because in India there many different forms of land and people farm on them, maybe on black soil, muddy, steppe farming, or plains. Our proposed tech is well suited for all.

6.3 OUR CONTRIBUTION

Our innovation focuses on the problems which directly affect farmers and the agriculture industry, including quality of the crops, quality of soil, safety measures, water measures, air quality measures, and threat detection. Our innovation uses both IoT and AI in coordination to monitor farm fields and the quality of the crops, rendering the data in visual form and location of the crop to the owner via SMS or our designed app. The AI also focuses on the best possible solutions needed to fix a certain issue. Fertility and quantity of the moisture of the soil are regularly checked through PH and moisture sensors which detect information and direct it to the user. Water in farm fields is regulated by the water level indicator. To prevent farm fires, smoke detectors are installed, to give early warning of any problem [4–7].

Our innovation involves the latest technology of AI and IoT. There is a microcontroller as a central hub and sensor modules. Each sensor with its specific tasks is dug into the farm at a particular range and connected to the microcontroller. The sensors then transmit the data to the central server and the hub transmits the data back to the user through a local wireless fidelity and user access them through our designed app or via SMS. The plant quality detection not only focuses on identification issues but also gives the best possible solutions for every particular case. Our water level indicators work on certain ranges so that the water required is minimal, thus avoiding water wastage. Major accidents caused by natural factors such as farm fires can be avoided by installing smoke detection sensors to give warning about the incident. Ultrasonic sensors are installed to detect threats and to prevent the rampage of crops caused by animals such as bulls, blue cow.

6.4 BASIC ALGORITHM FOR THE GIVEN SOLUTION

In Figure 6.1 we show two main sensors for water control, soil moisture and water level sensors, which are further combined with a rain sensor, for water irrigation perspective. With the help of soil sensor the acidic values of the soil are calculated and with the water level sensor the amount of water present in the soil is given. While the rain sensor gives an alert if there is rain in the field and the water level is already at the required level it will start auto recovery mode. If the water level is not high enough the water pump will be activated automatically and start supplying the water to the field until its maximum level is reached. That is designed and demonstrated in Figure 6.1 of module 1.

In Figure 6.2 we see the rain sensor, for water harvesting purposes, that is connected along with module 1 to generate a logic for irrigation and harvesting simultaneously. Irrigation is a major problem – at some places the water supply is very high and at a few places it does not reach properly to roots. The IoT based automated designed

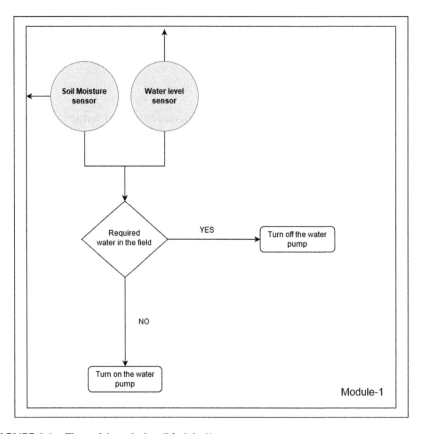

FIGURE 6.1 Flow of the solution (Module 1)

system, which is connected with the rain sensor and water level indicator, will provide the automated water supply with the nozzles placed at distant locations. An automated process of water harvesting will store the extra rainwater of rain for future use.

Figure 6.3 shows the mechanism designed to protect the crops from stray animals where ultrasonic sensors are used at different places and can generate an alarm if such conditions exist.

In Figure 6.4 is the automated rooftop designed for temperature sensitive crops where the installed temperature sensor informs the user about the current temperature of the field on a real-time basis, with an alert on specific conditions which are not suitable for the crops and vegetables. When an alert is generated an automated rooftop is activated to cover all the specified areas.

There are numerous sensors that are very much required for all data collection:

- Soil sensor
- Water level sensor
- Rain sensor

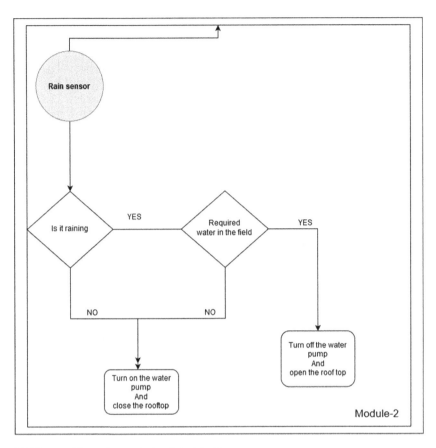

FIGURE 6.2   Flow of the solution (Module 2)

- Ultrasonic sensor
- Temperature senor

The use of these sensors is shown in Figure 6.5.

- The soil sensor and the moisture level sensor are integrated with all sensors that will measure the water level of the field and will take action as necessary. If the water level is high or if the moisture content is high, as per the threshold set, then the water pump will be turned off; if the water level is low and also the moisture of the field is also low then the pump will be directed to start.
- The rain sensor is integrated with soil and water sensor and if the water level is high and if it is raining then the rooftop will cover the field.
- The ultrasonic sensor will detect any prey under its threshold and it will warn the owner of the field so necessary action can be taken.
- The temperature senor will detect any sudden rise in temperature and will generate an alarm that will inform the owner that there may be fire in the field.

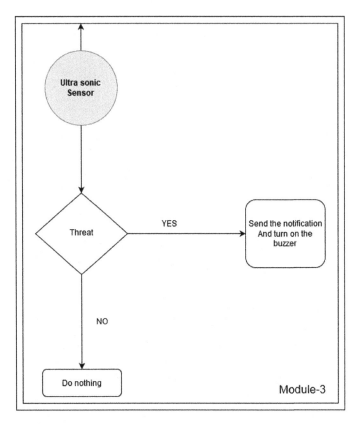

FIGURE 6.3   Flow of the solution (Module 3)

6.5 COMPONENTS USED

There are five main components/sensors used in this project. They are stated above and the description follows [10–13].

6.5.1 SOIL SENSOR

The sensor is utilized to gauge the level of water content (moisture) of soil. When the soil has a low amount of water, this gives an alert to maintain the proper amount of moisture required in the soil and reminds the client to supply the water to the plants and furthermore screens the moisture content of the soil.

6.5.2 WATER LEVEL SENSOR

This water level sensor module has sensor probe which helps in indication of 9 the water level in the soil. The probes send the information/ signal to the controller 10 which triggers the alarm, an indicator to notify the level of water.

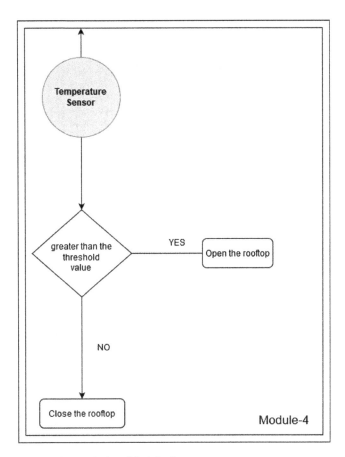

FIGURE 6.4 Flow of the solution (Module 4)

The water level depth detection sensor for Arduino has operating voltage DC3-5V and operating current less than 20mA. The sensor produces analog output signals according to the water pressure with its detection area of 40 × 16 mm. Easy to complete water to analog signal conversion and output analog values can be directly read on the Arduino development board.

6.5.3 RAIN SENSOR

A simple device for rain identification is used as a switch for the rooftop or cover to protect the crops from heavy and unneeded rain. This module has a rain board as well as the separate control board, power marker LED, and a customizable affectability with a potentiometer. This module permits you measure dampness by means of simple yield pins and it gives a computerized yield when a limit of dampness is exceeded. The module depends on the LM393 operation amp.

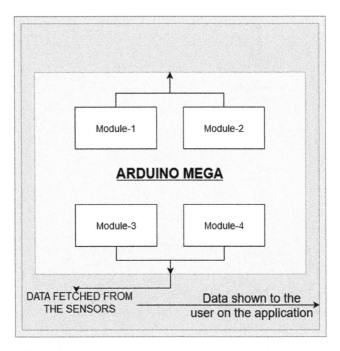

FIGURE 6.5   Final flow of the solution

6.5.4 ULTRASONIC SENSOR

Ultrasonic sensors in general give the results with the characteristics of acoustic waves with frequencies over the human perceptible range, often at around 40 kHz.

They regularly work by creating a high-recurrence beat of sound, and afterward accepting and assessing the properties of the reverberation pulse. Ultrasonic sensor module SRF-04 gives 2–400 cm non-contact estimation function, the running exactness can reach to 3 mm. The module incorporates ultrasonic transmitters, recipient, and control circuits in a single bundle.

SRF-04 transmits an ultrasonic pulse and calculates the time it takes to get the echo of the pulse back. The output received is in the form of pulse of variable width which is corresponding to the distance to the target. The SRF-04 module sends eight signals of 40 kHz by using an IO trigger for approximately 10μS high level signals and also detects whether its echo comes back or not. The time elapsed is the time duration from sending pulses from ultrasonic sensors to their returning. If the width of the pulse is measured in μS, then dividing by 58 will give you the distance in cm, i.e. μS/58 = cm, or dividing by 148 will give the distance in inches, i.e., μS/148 = inches.

6.5.5 TEMPERATURE AND HUMIDITY SENSOR

The humidity and temperature module, which produces aligned advanced yield. It can be used and interfaced with any microcontroller like Raspberry Pi, Arduino, to get prompt outcomes. This is a very low effort mugginess and temperature sensor

FIGURE 6.6 Soil sensor

FIGURE 6.7 Water level depth detection sensor

which produces high unwavering quality and long-haul strength. It utilizes a capacitive moisture sensor and a thermistor to quantify the encompassing air, and yields a computerized signal on the information pin (no simple information pins required). It is easy to utilize, and libraries and test codes are accessible for Arduino and Raspberry Pi.

This module makes it anything but difficult to associate the DHT11 sensor to an Arduino or microcontroller as it incorporates the draw-up resistor required to utilize the sensor. Just three associations are required to be made to utilize the sensor: Vcc, Ground, and Output. It has high unwavering quality and brilliant long-haul soundness, because of the elite advanced sign obtaining system and temperature and moistness detecting innovation.

6.6 PROPOSED SOLUTION

The proposed solution is an IoT-based system which is developed to automate the manual work done by farmers. It is an intelligent system which works upon artificial intelligence based algorithms and real-time data from different sensors. The proposed system will help farmers to know the condition of their crop and their farm in real time.

The yellow rust problem of wheat is resolved by applying the image processing algorithms. The detection of yellow rust and its level of growth in the wheat are provided to the farmers to take precautionary measures. Yellow rust generally occurs

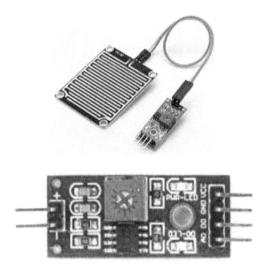

FIGURE 6.8   Rain drop sensor

FIGURE 6.9   Ultrasonic sensor

in the winter season and continues into the early stage of the spring season. The yellow color band of leaves starts converting into oranges stripes on adult plants.

The data, like soil moisture, water level, temperature, humidity, are collected from the farm using different sensors in real time. The real-time data is passed to the algorithm which is functional upon Raspberry Pi. The AI algorithm takes decisions according to the input data provided and the decision is passed back to the electronic devices like DC motor (for movable sheds), LED lights, and many more to perform actions to save the crops and supply required water and fertilizers in the farm. The proposed system also uses image recognition to detect yellow rust on the crop and alert the farmer about that defect.

The proposed system is connected to an android based app designed in flutter to help the farmer know the real-time data on the farm. The farmer can also use the app

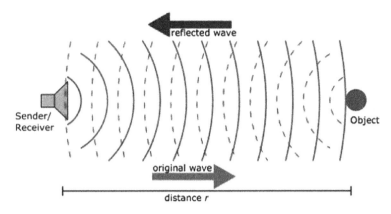

FIGURE 6.10   Working principle

FIGURE 6.11   DHT 11 sensor

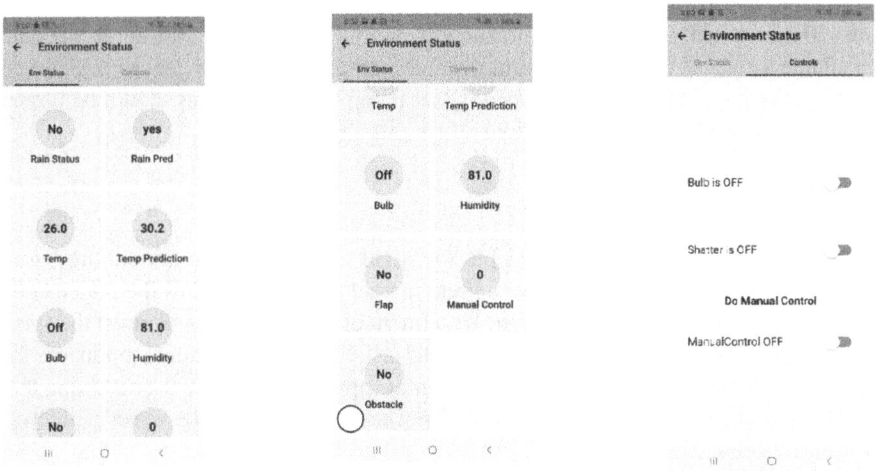

FIGURE 6.12   Statistical evaluations of different parameters

to control motors, shed, LED lights, and other electronic devices, which can help the farm have manual control over the AI based algorithm. The proposed system uses image processing to detect the condition of crops and uses sensors to help the farmer to take appropriate decisions. The proposed system also handles excessive rain by using an automated roof top system for ground cover crops. The yellow rust detection process is implemented using Open-CV in Python. We tried to filter a specific color of yellow. In order to filter like this, image colors are converted to HSV, which is "Hue Saturation Value." Hue is the color, saturation provides the strength of the color, and value is taken for light. This will try to identify a more specific color, on the basis of hue and saturation ranges, with a variance of value. The converted HSV is passed on to create a mask using erode and dilate. As a result, we are able to find the yellow rust in the sample frame.

This is only an example, with yellow color as the target. Whatever we will capture and see will be anything that is between our ranges here, basically (21,70,80) to (61,255,255). This is for yellow. We are working with the range of colors and HSV is working with the color range., To figure the HSV range, the best method is to just trial and error. There are built in methods to Open-CV to convert BGR to HSV. If you wanted to pick just a single color, then the BGR to HSV would be great to use.

In Figure 1.13 the process of image processing is explained. Where the problem of yellow results in the wheat is considered, high resolutions cameras are required to capture the images of the crops and these are converted into HSV. The assigned color range is to be tracked with the color range of HSV images. Masking of the process and then identifying the error is providing the results of the detection of yellow rust.

6.7 RESULTS

In this section, there are seven cases taken to decide the parameters of the proposed system (see Table 6.1). There were different inputs given to rain module sensor, soil moisture sensor, and yellow rust detector to find out how the output varies in different conditions for the wheat crop. The study of wheat crop tells that it needs water level of 450 mm to 650 mm and soil moisture of 35 to 40 percent. According to the data available for wheat crop, the algorithm was designed to save the wheat crop. All seven outputs helped us to know when shed, water pump, fertilizer pump should be turned on and off according to the conditions of the farm and the crop.

6.8 CONCLUSION

The proposed solution is very feasible for small farms that have agriculture based on ground cover crops. It will have a great impact on the farmers as the data can be processed and will be provided for the betterment of the corps and will boost the agro-industry sectors. The basic system is easy to install and is very much applicable for all the ground crops. The data are shown on the application on the user's phone that is much multilingual and very user friendly. Therefore, the solution is very effective for the small field farmer owners.

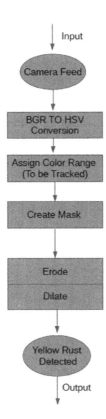

FIGURE 6.13 Statistical evaluations of different parameters

TABLE 6.1
Overall compilation results of proposed solution

	INPUT				OUTPUT			
Cases	Rain	Soil moisture	Water level	Yellow rust	SHED	Water pump	Fertilizer pump	Alert
Case 1	Yes	10%	2 mm	No	Up	Off	Off	High
Case 2	No	20%	4 mm	No	Up	On	Off	Low
Case 3	No	40%	5 mm	Yes	Down	Off	On	No
Case 4	Yes	50%	5.5 mm	No	Down	Off	Off	No
Case 5	Yes	60%	6 mm	No	Down	Off	Off	No
Case 6	Yes	80%	8 mm	Yes	Down	Off	Off	High
Case 7	Yes	100%	10 mm	Yes	Down	Off	Off	Very High

BIBLIOGRAPHY

[1] SciForce Smart Farming the Future of Agriculture, 14 January 2019, https://medium. com/sciforce/sciforce-who-we-are-ebe1dbac447a

[2] A. Patil, M. Beldar, A. Naik, and S. Deshpande, "Smart farming using Arduino and data mining," *3rd International Conference on Computing for Sustainable Global Development*, New Delhi: INDIACom, 2016, pp. 1913–1917.

[3] N. Gondchawar and R. S. Kawitkar, "IoT based smart agriculture," *International Journal of Advanced Research in Computer and Communication Engineering*, 5(6), ISSN (Online) 2278-1021 ISSN (Print) 2319 5940, June 2016.

[4] P. Rajalakshmi and S. D. Mahalakshmi, "IOT based crop-field monitoring and irrigation automation," 10th International Conference on Intelligent Systems and Control (ISCO), 7–8 January 2016, published in IEEE Xplore, November 2016.

[5] T. Baranwal and N. P. K. Pateriya, "Development of IoT based smart security and monitoring dev ices for agriculture," 6th International Conference – Cloud System and Big Data Engineering, 978-1-4673-8203-8/16, 2016 IEEE.

[6] N. Khan, G. Medlock, S. Graves, and S. Anwar, *GPS Guided Autonomous Navigation of a Small Agricultural Robot with Automated Fertilizing System*, SAE Technical Paper, Univ of California Berkeley, 2018, doi:10.4271/2018-01-0031

[7] K. V. d. Oliveira, H. M. Esgalha Castelli, S. José Montebeller, and T. G. Prado Avancini, "Wireless sensor network for smart agriculture using ZigBee Protocol," 2017 IEEE First Summer School on Smart Cities (S3C), Natal, 2017.

[8] R. Chandana, S. A. K. Jilani, and S. Javeed Hussain, "Smart surveillance system using Thing Speak and Raspberry Pi," *International Journal of Advanced Research in Computer and Communication Engineering*, 4(7), 2015, pp. 214-219.

[9] D. A. Visan, I. B. Cioc, A. L. Lita, and S. Opera, "GSM based remote control for distributed systems," *ISSE, 2016 39th International Spring Seminar*, September 2016, pp. 2161–2064.

[10] H. Saini, A. Thakur, S. Ahuja, N. Sabharwal, and N. Kumar, "Arduino base automatic wireless weather station with remote graphical application and alerts," *SPIN, 2016 3rd International Conference*, September 2016, 978-1-673-9197-9.

[11] N. Putjaika, S. Phusae, A. Shen-Im, P. Phunchongharn, and K. Akkarajitsakul, "A control system in an intelligent farming by using Arduino Technology," *ICT-ISPC, 2016 Fifth ICT International*, 2016, p.p. 51–61.

[12] V. Naik, S. Pushpa Bai, P. Rajesh. and M. A. Naik, "IoT based green house monitoring system," *International Journal of Electronics and Communication Engineering & Technology (IJECET)*, 6(6), 2015, pp. 45–47.

[13] C. Kavitha, C. Ramesh Gorrepotu, and N. Swaro, "advanced domestic alarms with IoT," *International Journal of Electronics and Communication Engineering and Technology*, 7(5), 2016, pp. 77–85.

7 Efficient Green Solution for a Balanced Energy Consumption and Delay in the IoT-Fog-Cloud Computing

Neeraj Dahiya, Surjeet Dalal, and Vivek Jaglan

CONTENTS

7.1   INTRODUCTION

The cloud offers an efficient computing model where resources such as online applications, computing power, storage, and network infrastructure can be shared as services through the internet. In recent years, cloud systems have provided a solution to the IoT, the digital interconnection of everyday objects with the internet. Currently, IoT applications are increasing in various fields. However, the number of devices connected to the internet worldwide (therefore accessing cloud services) is in continuous growth. This means that the communication latency and the power consumed during the communications with the cloud reduce the expected advantages of this technology. The energy consumption of all devices in the home-based fog computing environment and the energy-delay trade-off in different levels of the fog-cloud interaction have been highlighted widely in research, using tools from stochastic optimization [6]. Since the devices/sensors are energy-constrained, works like [7] have focused on some issues like residual battery lifetime. The energy-characteristics of communications in these devices and the world of IoT are described in Figure 7.1.

Our models formulate the energy consumption and the delay for the entire execution cost rather than focusing only on the devices. Our main contribution is to investigate the problem of energy consumption of the fog computing in the context of IoT applications, and propose a balanced energy-delay solution based on an evolutionary

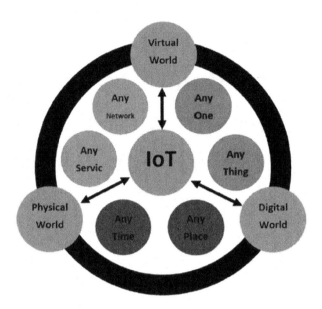

FIGURE 7.1   IoT scenario

algorithms approach. As we can see, cloud computing is becoming the overarching internet approach for information storage, retrieval, and management, and IoT devices become the major outlets of service applications. This means that the key task for the next generation network is the successful integration of cloud computing and IoT devices. Compared to cloud computing, fog computing can provide enhanced service quality with much increased data rate and reduced service bandwidth cost [5]. However, like IoT devices, fog nodes are resource constraints as well. Therefore they need to contact the cloud often as some computations may require an amount of resources that cannot be handled singly by the fog node. When we can fulfill as many requests as possible for end users we have much fewer requests to be sent to the cloud and less bandwidth cost. The greenhouse industry is the fastest growing sector. Greenhouses separate crops so they get a controlled environment and fertilization. The main advantage of the greenhouse is we can produce many crops at a time by manipulating environmental conditions as the different crops require. This allows the farmer to improve the cultivation in a way the plants need. It leads to higher crop yield, prolonged production period, better quality, and less use of protective chemicals. In a moderate climate zone, energy is needed, whereas in arid zones, the cooling and availability of water is a major concern. The use of materials and energy as well as crop yield and quality can be influenced by operating the adjustable components of the greenhouse, such as heating and cooling inputs, window opening, drip irrigation, screening, and CO_2 dosage. Hence, it can be expected that the way these controls are operated influences the final economic result [9]. According to the requirement of the crops the threshold will be set, so if any environmental conditions like temperature, soil conditions, and humidity, go below or above the threshold value, then the changing in parameters are monitored simultaneously and all the data will be transmitted to farmers. The farmer will then take the controlling decision and send it to the system. The system will run the actuator and control the parameter.

Green building is a rational certainty for sustainable development, and is the future trend of architecture. It achieves its goals with the technologies of Internet of Things (IoT) and cloud computing, with the green building energy efficiency standards and green-building-related technical codes. A building IoT system has the ability to make all resources, such as people, environments, sensors, and physical devices, into an integrated system [1]. There are a lot of automatic control systems used in buildings. Especially in China, equipment systems and energy consumption monitoring platforms are encouraged by government to apply to public buildings or commercial buildings. They will cover more than 20 billion square meters in 2020 [11]. With the development of the social economy, the functions of high-rise constructions become more complex and lead to two problems. One is that more manpower and material resources are needed to maintain building equipment. The other is how to improve the performance of data storage and concurrent communication. In order to solve these two problems, it is essential to set up a rational building operation system for green buildings [13, 1]. There is a tendency for buildings to be constructed with a building integrated system, which usually provides lots of monitoring devices and control modules. However, a traditional building integrated system cannot offer appropriate service when the main purpose of the building is changed. In addition, the abilities of various end user access and device communication require good performance inf concurrent communication, which should be considered at the beginning of system design. To build a new building control system for green buildings, this chapter presents a building management cloud platform based on IoT and cloud computing. All functions are modularized for the building management cloud platform. The building management services are also modularized.

Overall, we want to corroborate that further research into peer-to-peer fog computing is worth the effort and large-scale networks should adapt peer-to-peer fog computing architectures to extend and improve traditional fog computing. The P2PFOG model proposed in this research is motivated by IoT service requirements, user demands, and technology progress, which to the best of our knowledge has not been addressed before in a coordinated fashion. As well as for the traditional cloud computing paradigm, we think that also IoT cloud should look at federation in order to address new business scenarios. This statement is motivated by the rapid evolution of the container virtualization technology that is more suitable for IoT devices than the traditional hypervisor virtualization technology. In fact, the container virtualization technology offers better performance on IoT devices and enables virtual sensor migration. Energy efficiency is one of the major keywords in the cloud computing literature [4], including two main aspects that are generally uncorrelated: energy costs-saving and energy sustainability [5]. In addition, IoT scenarios require proper energy efficiency management strategies. In this chapter, we propose a strategy to improve the energy sustainability and of a whole federated IoT cloud ecosystem. In particular, we refer to a use-case driven federated IoT cloud scenario, in which each IoT cloud provider has several functional requirements and is interconnected with other providers in a federated environment by means of a cloud broker. We define the concept of IoT cloud federation as a mesh of IoT cloud providers that are interconnected to provide a universal decentralized sensing and actuating environment where everything is driven by constraints and agreements in a ubiquitous infrastructure. IoT cloud federation is mainly based on the container virtualization technology that allows providers to deploy computational resources (i.e., containers) on smart sensors in order to run IoT services, and to move (or migrate) them into other smart sensors belonging to the federation. This approach allows IoT cloud providers to take advantage of a resilient IoT virtualization infrastructure including both virtual sensors and actuators. On this basis, an IoT cloud provider has to be able to cooperate with other ones, thanks to the intermediation of a broker, in order to achieve an efficient energy management.

The "smartness" of a city means its ability to bring together all its resources, to effectively and seamlessly achieve the goals and fulfill the purposes it has set itself. In other words, it describes how well all the different city systems, and the people, organizations, finances, facilities, and infrastructure involved in each of them work. Smart can be defined as an implicit or explicit ambition of a city to improve its economic, social, and environmental standards. The concept of smartness in terms of performance is highly relevant to technologically implementable solutions. In many cases, if there is some form of ICT which is present in a city, the city or its activity is considered "smart". ICT devices and services are only an enabler or purveyor which allows the "smartness" to percolate throughout a system. Just by having a personal computer or smart phone does not guarantee "smartness" or intelligence. Specifically, the International Organization for Standardization has recently released a report on this issue [7]. Embedded systems have always been held to a higher reliability and predictability standard than general-purpose computing. Consumers do not expect their TV to crash and reboot. They have come to count on highly reliable cars, and the use of computer controllers has dramatically improved both the reliability and efficiency of cars. In the transition to CPS, this expectation of reliability will only increase and the same is demonstrated in Figure 7.2.

In fact, without improved reliability and predictability, CPS will not be deployed into such applications as traffic control, automotive safety, and health care. Smart city research must also examine the roles of various cyber-physical systems. Today's cities have not only physical, but also virtual/cyber infrastructures. Cyber-physical

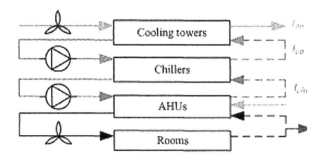

FIGURE 7.2 Flow and temperature

systems connect the virtual world with the physical world with the help of embedded systems and data communication systems. Lee et al. write the following:

A cyber-physical system is an integration of computation with physical processes whose behavior is defined by both cyber and physical parts of the system. Embedded computers and networks monitor and control the physical processes, usually with feedback loops where physical processes affect computations and vice versa.

[8]

We consider the fog architecture in what follows. We suppose that the cloud has one DC attached to the network core. The network at the edge includes several fog nodes called fog instances. These fog instances can cooperate (by sharing network, computational capacity, and storage) to provide an optimal service for IoT objects. Considering this settled environment, we define the performance metrics. In our genetic algorithm each object (respective Fog instance) is encoded by its respective index. A chromosome corresponds to a distribution of all the assignment objects-fog instances possible. The fitness of a chromosome represents the total amount of the energy consumed by the fog instances used in the solution, with regard to the same constraints described before. According to the selected assignment object fog instance, it will determine the performance of our solution. We start by encoding the objects and FIs, so that we can generate a population including all the genes (all the possible assignments object-fog). An IoT system includes an IoT input device, IoT cloud, and IoT output device. The IoT input devices will have various input modules attached that can sense what is going on in the environment (e.g., light, temperature, humidity, soil control, CO_2 control) and be able to upload data into the cloud. We will be using IoT cloud service. Each IoT output device will be able to download data from the cloud and will have various output modules attached to display data (e.g., LEDs, piezo buzzer, mini-printer) or react in some way (e.g., servo, relay for controlling appliances) All the sensors collect the current situation of the greenhouse and send data to the user. The user will check all information sent by the system and send a controlling algorithm to actuator via the cloud so that actuator will control the changes in parameters and the building management cloud platform adopts browser server structure. It not only realizes overall building environmental monitoring, but also provides management, control, data processing, and other services. Users can operate and maintain the building simply by clicking the modules. An intelligent device level is the basis for comprehensive perception of the cloud platform, which mainly includes intelligent hardware such as environmental node, equipment control node, energy consumption measuring node, and networking relay. The function of environmental node is to collect building environmental data of temperature, carbon dioxide, illumination, humidity, formaldehyde, ammonia, benzene, and solid particulate matter. Equipment control nodes provide convenience that electrical equipment can be controlled easily by the building management cloud platform. The energy consumption node is installed in an electricity meter or electrical equipment for measuring voltage, current, and power. All these nodes interact with the networking module to the cloud platform at the networking level. Based on the cloud database, the building management cloud platform offers green building management like API, ubiquitous access, registration, controlling, and optimization. Services covering monitoring, controlling, display, data analysis, alarming, and information pushing are integrated in the modularized service level. They are customized according to the requirements of the user by simply dragging the modules on web browser. The main procedure processes data from other intelligent devices such as environmental node, energy consumption node, equipment control node. It realizes Wi-Fi connectivity inspection, web server status inspection, TCP client status inspection, data collection, and data processing. Cloud computing is the practice of using remote servers on the internet to manage, store, and process data instead of using a personal computer. Cloud computing is a general term that is better divided into three categories: infrastructure-as-a-service, platform-as-a-service, and software-as-a-service. IaaS (or utility computing) follows a traditional utilities model, providing servers and storage on demand with the consumer paying accordingly. PaaS allows for the construction of applications within a provider's framework. SaaS enables customers to use an application on demand via a browser. About measuring and modeling cloud computing clusters or specific platforms, in [11] the authors present an experimental method finalized to the configuration, monitoring, and maintenance of a Raspberry Pi cloud cluster consisting of 500 nodes. In [12] a power model is presented, named POWERPI, to estimate the power consumption based on CPU and network utilization. The authors show it is possible to reduce power consumption by optimizing the software running on the platform, thus to derive possible power saving strategies. By starting from this issue, an IoT cloud provider is able to ask for external resources (e.g., virtual sensors and actuators) to other providers for multiple reasons, including resource enlargement, resource optimization, costs reduction, security, QoS, deployment of distributed services, fault-tolerance, energy efficiency (for reducing energy costs and/or for energy sustainability), and so on. This is possible by means of the container virtualization on IoT devices that allows to migrate resources or services (virtual sensors and actuators) from one IoT cloud to another. In this chapter, we focus on energy sustainability in IoT cloud federation. Our approach allows the federated ecosystem to implement specific policies to dynamically select the best destination where to allocate/migrate virtual sensors and actuators. Our approach leverages the dynamic deployment of containers to run services with good performances and reduced startup time [15]. Although this means assuming that costs for the hosting service are strongly related to the availability and the effective energy consumption at the specific cloud sensor node, our strategy is focused on improving energy sustainability for the whole federated cloud ecosystem. Cloud computing is defined as a "model for enabling convenient, on-demand access to a virtualized shared pool of computing resources (e.g., servers, storage, applications, and services) that can be rapidly provisioned and released with minimal management effort" [6]. The 2010 Davos World Economic Forum highlighted the benefits of cloud technologies compared to traditional systems: (1) fast and easy deployment with no need for extensive IT experiences, or build large IT backbone infrastructure, (2) scalability where organizations can scale up or down their IT infrastructure according to their consumption needs, and (3) distribution of cost through multi-tenancy and pay-as-you-go models [7] [8]. In particular, we branch the monitored environment in several

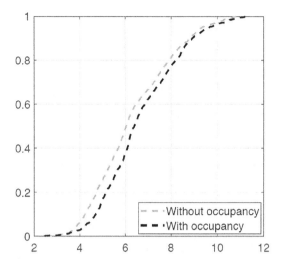

FIGURE 7.3 Energy consumption in IoT devices

sub-environments where each one represents the minimum coverage area for measurement that is determined on the basis of a specific use-case.

We use virtual sensing and actuating techniques to deploy resources and/or services in the destination IoT device. A broker that receives a request from an IoT cloud provider identifies the use-case in terms of phenomena, accuracy in the size of the sub environment, and activities. Therefore, it starts to look for external providers belonging to the federation on the basis of this use-case, in order to determine several possible available IoT clouds. At the subsequent step, the IoT cloud requesting the hosting service receives from the broker an input dataset including the energy parameters of external federated clouds. This dataset is mainly expressed in terms of the power consumption requested to run the virtual sensor. Thus, the IoT cloud runs an algorithm in order to determine in which federated IoT cloud this virtual sensor should be deployed. However, since the requesting IoT cloud does not want to expose its businesses, it sends to the broker a data stream which contains only values to solve the request. The fog computing scenario is a basic three-layer fog computing architecture, which extends the cloud architecture by adding an intermediate fog layer [4] as illustrated in Figure 7.4. This fog layer responds to clients' requests and sends requests to the cloud if needed. In this scenario, fog nodes are pre-cache units that are empty at the beginning. Every time a fog node receives a file request from a client, it first checks if it contains the requested file. If the fog node contains the file, it responds to the client node with the requested file. Otherwise, the fog node starts a lookup request for the file from the cloud.

The cloud then responds to the fog node with the requested file. Now, as the fog node receives the file, it caches the file and sends it to the client. It tries to search for the file in a static amount of time. Every P2P lookup has a maximum time for searching the files. If one of the routed fog nodes contains the file, it replies with the file to the requesting fog node. Again the file is cached in the fog node before being

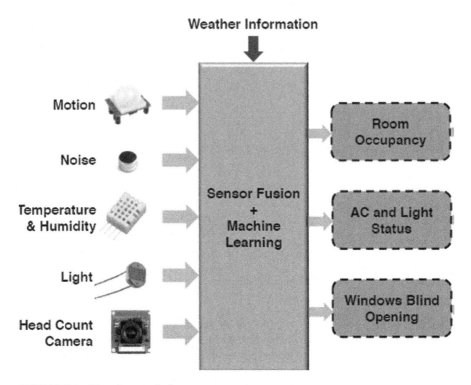

FIGURE 7.4   Use of sensor fusion

sent to the client. But there is still the chance that none of the routed fog nodes will respond with the file, either because the P2P lookup time is over, or because the file has never been requested before, so it is not cached in any fog node. In this case the requesting fog node starts a new lookup request. This time it directly requests the file from the cloud. The cloud responds with the file, and the fog node caches the file, before sending it to the client.

Big data management is essentially the core software layer that ultimately drives the BMS through big data analytics that includes prediction (e.g., predictive maintenance), model building, complexity mapping, and visualization. Big data management aids the dynamic management of energy consumption via monitoring and analyzing the energy-related activities to minimize unintended energy and water consumptions. It is essentially the process of examining the large dataset obtained through the IoT. Through this process, the building manager is able to identify information on human activity, weather conditions, the micro-climate within the building, and related energy wastage. To do so, algorithms need to be designed that can accurately extract the inter correlation between the load consumption, human occupancy and movement activity, energy wastage, renewable energy generation, weather conditions, and hence design effective models from real-time large-scale datasets in order to perform predictive maintenance.

7.2 AN ARCHITECTURE FOR ASSESSING GREEN SOLUTIONS USING IOT

We have showcased how various advances are enabling the assessment of safety of vehicles, roads, and drivers. The objective of this section is to introduce a novel and adaptive IoT architecture that enables the assessment of safety in a city's road network. We elaborate on the assessment elements and how they can be used to synthesize a single, meaningful indicator for safety. We also describe the architecture components and their interrelationships, including a robust computational core for safety assessment. While an objective of the work is to realize a cost-effective solution for dynamic assessment, the purpose of the setup described above was designed for both exploratory and validation purposes. Where the sensing elements were able to connect to the cloud repository directly, e.g., as in the case of the box dongle, or the smartphone, the connection was allowed. Otherwise, a connection was established indirectly through another device, e.g., the logger through the PC laptop or Scan Tool's dongle through the smartphone. Overload or congestion is one of the major challenges in WLAN-based IoT as well as in WWAN-based IoT. In this regard, Liu et al. [13] proposed an offset listen interval algorithm to alleviate network contention and delay. The algorithm ensures that M2M traffic be spread evenly in WLAN beacon periods with calculated offsets; access points make wake-up schedules in such a way that a minimum number of devices wake up during a certain beacon period. However, we think that there needs to be some enhancement to assure different delay bounds for delay-sensitive IoT devices for emergency applications. Lin et al. [14] suggested a deep sleep mechanism that enhances IEEE 802.11 PSM by granting a higher channel access priority for low-energy devices dynamically. The mechanism resolves the overhearing or congestion problem as device wake-up time is randomly deferred. IoT devices have to wake up after a certain amount of time and then an IoT gateway will transmit all buffered packets to all the devices. An analysis of energy and delay under two contention resolution mechanisms, i.e., frame slotted Aloha and tree splitting contention resolution, is presented in [11]. Based on the optimal packet length, the mechanisms assure energy efficiency and minimum delay for M2M networks. If there is a fault in the fan, the air flow rate obtained from the lookup table based on the fan frequency would deviate from the actual air flow rate. Using the improper airflow as an input to the cooling tower model would cause estimated parameters to deviate from their normal ranges and cause the fault detected. Power conservation optimization is a challenging area that requires novel approaches to solve the underlying issues. Power is a vital resource for battery-operated systems. Recent technological advancements have paved the way for miniaturization in terms of size and weight of such systems. However, power consumption remains a dominant issue in the design process. Both aspects of a system, i.e., software and hardware, are to be considered in evaluating the true power consumption. Typically, the hardware part is considered to be more resource hungry. Among the issues within context awareness, power conservation is of prime importance. Since a context-aware system can be mobile, and it relies on the internal sensors present on the device as well as the external sensors present in the environment, there is a need to conserve power. Furthermore, the context recognition is itself a power-consuming task. Once the context is gathered, it must be processed, and processing consumes power. This technique requires a priori information to effectively predict when to poll the sensors, and when to process the context if continuous data are provided from the sensors. If the classification of the context is carried out as a predictive algorithm, it would support the power awareness task. Otherwise the system would use two different learning algorithms, i.e., one for classification of the context and other for sensor polling.

7.3 CONCLUSION

Typical ecosystem architecture illustrates the sensor nodes communication between smart objects or with in the objects using the different layers of the protocols like MAC layer, network layer, processing layer, service layer, and business layer. Although WSNs are a truly cross-disciplinary research topic, one major challenge is to enable network-wide communication. Because batteries cannot be replaced, communication protocols ought to be as energy efficient as possible. This poses a challenge as it implies turning off the radio chips most of the time. Unlike traditional wireless networks, source and destination nodes are not within communication range most of the time, and intermediate nodes need to relay the message. In this multi-hop scheme, new problems arise such as handling the increased load due to the relay process, or finding a suitable multi-hop path between source and destination.

In this chapter, the Internet of Things has been described as a network of networks and we talked about its history, three visions, and developments. The smart grid is one of the most important applications of IoT and its architecture and elements are discussed. Fault detection of these systems is difficult since these systems are large in scale, consisting of many coupling subsystems, building and equipment dependent, and working under time-varying conditions. In reality, such systems will run for weeks or even months, meaning that the difference margin will be much bigger. However, even for simulation time, compared to realistic IoT scenarios, our proposed P2P fog model provides better bandwidth throughput compared to the cloud computing and fog computing. In this work, we presented a model to investigate the power consumption and delay in fog-cloud computing and in the traditional cloud computing. Then, we proposed a solution inspired from evolutionary algorithms approach to resolve the trade-off problem. Simulations and numerical results have shown the practical relevance of our approach with a large number of real-time, low latency IoT applications.

BIBLIOGRAPHY

[1] C. J. Palmer, S. D. King, R. R. Cuadrado, E. Perez, M. Baum, and A. L. Ager, Evaluation of the MRL diagnostics dengue fever virus IgM capture ELISA and the PanBio rapid immunochromatographic test for diagnosis of dengue fever in Jamaica, *Journal of Clinical Microbiology,* 37 (1999), 1600–1601.

[2] J. Ni, K. Zhang, X. Lin, X.S. Shen, Securing fog computing for internet of things applications: Challenges and solutions. *IEEE Commun. Surv. Tutorials.* 20 (2017), 601–628.

[3] N. Dahiya, Chemical reaction-based optimization algorithm for solving clustering problems, in *Natural Computing for Unsupervised Learning*, Cham: Springer, 2019, pp. 147–162.

[4] P. Neuzil, J. Pipper, and T. M. Hsieh, Disposable real-time microPCR device: Lab-on-a chip at a low cost, *Mol. Biosyst.* 2 (2006) 292–298.

[5] O. J. Brady, P. W. Gething, S. Bhatt, J. P. Messina, J. S. Brownstein, A. G. Hoen, et al., Refining the global spatial limits of dengue virus transmission by evidence-based consensus, *PLoS Negl. Trop. Dis.* 6 (2012) 1–15 e1760.

[6] B. Seth, S. Dalal, and R. Kumar, Securing bioinformatics cloud for big data: Budding buzzword or a glance of the future, in R. Kumar and U. Wiil (eds), *Recent Advances in Computational Intelligence. Studies in Computational Intelligence*, vol. 823, Cham: Springer, 2019. https://doi.org/10.1007/978-3-030-12500-4_8.

[7] J. Gómez-Luna, H. Endt, W. Stechele, J. M. González-Linares, J. I. Benavides, and N. Guil, Egomotion compensation and moving objects detection algorithm on GPU, in *Proceedings of the Conference Applications, Tools and Techniques on the Road to Exascale Computing*, Ghent ParCo, 2011, pp. 183–190.

[8] V. Pham, P. Vo, H. T. Vu, and H. B. Le, GPU implementation of extended gaussian mixture model for background subtraction, in *Proceedings of the IEEE RIVF International Conference on Computing & Communication Technologies, Research, Innovation, and Vision for the Future (RIVF)*, Hanoi: IEEE, 1–4 November 2010, pp. 1–4.

[9] B. Seth, S. Dalal, and R. Kumar, Hybrid homomorphic encryption scheme for secure cloud data storage, in R. Kumar and U. Wiil (eds) *Recent Advances in Computational Intelligence. Studies in Computational Intelligence*, vol 823, Cham: Springer, 2019. https://doi.org/10.1007/978-3-030-12500-4_5.

[10] D. Blalock, S. Madden, and J. Guttag, Sprintz: Time series compression for the internet of things, *Proceedings ACM Interactive Mobile Wearable Ubiquitous Technology,* 2(3), 2018, pp. 93:1–93:23, doi: 10.1145/3264903.

[11] B. Seth and S. Dalal, Analytical assessment of security mechanisms of cloud environment, in K. Saeed, N. Chaki, B. Pati, S. Bakshi, and D. Mohapatra (eds), *Progress in Advanced Computing and Intelligent Engineering. Advances in Intelligent Systems and Computing*, vol 563, Singapore: Springer, 2018. https://doi.org/10.1007/978-981-10-6872-0_20.

[12] C. Rocha, Developments in national policies for food and nutrition security in Brazil. *Dev. Policy Rev.* 27, 2009, 51–66.

[13] D. Ahluwalia, Public distribution of food in India: Coverage, targeting and leakages, *Food Policy*, 18(1), 1993, pp. 33–54.

[14] Arab Academy for Science, Technology and Maritime Transport, United Nations International Strategy for Disaster Reduction, World Bank, Workshop Summary Report, Regional Workshop on disaster risk reduction and climate change, Cairo 21–23 November 2009.

[15] African Union/NEPAD/African Development Bank/African Development Fund/World Bank/United Nations, *International Strategy for Disaster Reduction, Africa Regional Strategy for Disaster Risk Reduction*, 2004.

8 Mobility Management in Green IoT

Neeraj Dahiya, Surjeet Dalal, and Vivek Jaglan

CONTENTS

8.1 INTRODUCTION

Changing the overall essentialness structure may be a dangerous tradeoff between advancing money-related reality and safeguarding the world. Against this foundation, reducing CO_2 transmissions is of essential importance so as to urge a practical development pathway. The Paris Agreement recommends restricting the overall temperature increase to 21°C. It has been observed that a useful essential system change requires solidified undertakings of innovative development, money-related improvement, technique intervention, and social change. Among such changes, a couple of perspectives are particularly reassuring and subsequently have attracted great interest, for example low-carbon power methodology. Additionally, the event of a decentralized age and limit in centrality structures has confused the division between standard creators and purchasers, giving rise to "prosumers." We are at a fundamental point within the qualification in our centrality system from generally isolated vitality storerooms and direct effectively chains into interconnected complex structures with associate parts and adornments. Along these lines, the conspicuous affirmation of open requesting and potential responses for this from a general point of view significant change presents centrality authorities with exceptional possibilities.

This far-reaching overview and impact appraisal will investigate and fundamentally explore electronic developments.

8.2 ELECTRONIC PHYSICAL STRUCTURES

The objective of this chapter is conceptualizing the various updates of electronic advances and their impact on essential structures with unequivocal focus on regular sensibility and money-related legitimacy. It gives an outline of crucial innovations and related applications. Despite the way that pushed upgrades are referenced in various plans, at long last these different advances are consistently combined with one another in unequivocal applications. For instance, an advanced metering structure (AMI) may be a huge wellspring of giant data in essential systems; in any case, evaluation of enormous data might be formed through artificial intelligence (AI) and machine learning (ML). So, additionally, the Internet of Things (IoT) can lead to a situation where data sharing through the semantic web is pressing to form a virtual outline of physical substances. The possibility of cyber physical systems (CPS) has been raised.

CPS are thus the strategy of related PCs and physical systems both on a level plane (inside a body and PC independently) and vertically (blend between a physical system and PC). In this chapter, CPS are delineated as co-produced interfacing structures of physical and computational parts, while the authentic frameworks are recommended as subsystems. CPS are expected to form a virtual depiction (the catch) of real substances (physical space) to seek perfect responses for authentic issues by seeing approaches within the web. In like manner, AI is often used alongside CPS. This blend of quick CPS in essential structures cannot simply change their structure rule and action framework. In any case, CPS had teething issues and picked models and their impact are going to be analyzed in this chapter. Regardless, a uniform IEA report points out that purposefully interconnected structures could on an astoundingly focal level change the present industry. The US Department of Energy's Clean Energy Smart Manufacturing Innovation Institute (CESMII) supports advances in smart sensors and digital process controls. Hence, it's essential to start a cautious discussion of how CPS types of progress (for instance IoT, AMI, ML got alongside AI) are often applied within the significance structure reach and improve its budgetary issues, sensibility, adaptability and security, while catalyzing decarbonization attempts. The colossal advances in computational power alongside ML methodology are currently enabling unequivocal AI at present. This chapter tries to introduce an impact assessment of CPS upgrades for the potential in significance structures, while concentrating on pulling in progresses, potential applications, impact on energy framework money related viewpoints and standard reasonableness similarly as centrality security. The second section gives a survey of the leading edge of the uncommon development structures for immensity frameworks. The third section explores how CPS are affecting these models utilizing a few of instances of CPS applications. The fourth section reviews the effect of eminent CPS on the budgetary deciding ability and adaptableness of centrality structures and deducts philosophy suggestion from this cash-related assessment.

8.3 EXISTING TURNS OF EVENTS

Supervisory control and data acquisition (SCADA) structures have for quite a while been utilized to screen and control essential water foundation. Despite standard SCADA structures, there has been an unremitting effect within the improvement of wireless sensor networks (WSNs) for water assets by the administrators. While these advancements have made stunning advances around connecting with and control of water frameworks, a nonattendance of as far as possible plans has controlled the structure scale and the administrators of watersheds. Here, we audit existing creative reactions for water structure checking and control, and describe how open tornado moves the highest level by giving the central open source.

8.4 REMOTE SENSOR STRUCTURES

The previous decade has seen a huge reduction within the expense and forced utilization of remote contraption; utilizing these advances, WSNs have opened new areas in trademarks, with applications running from biodiversity checking, woods fire district, accuracy making, and major thriving surveys. WSNs are perfect for unnecessary effort, low-force, and low-reinforce applications, making them suitable for the checking of colossal water frameworks like conductors and watersheds. WSNs are applied to enormous accomplishments in applications partners from flood checking to consistent water control; regardless, stream use is routinely foundation or restrictive, understanding a nonattendance of discoverability, constrained interoperability, and duplication of exertion. Inside the water sciences, flood checking addresses an especially immense application area for WSNs. While there have been endeavors to create the goals of existing heritage flood affirmation systems, scarcely any groups have shaped and sent their own flood checking WSNs. Hughes et al. (2008) depict a 15-focus riverine flood seeing WSN within the UK, which interfaces with remote models, performs near estimation, and sends neighborhood express flood advice. Other riverine flood watching structures consolidate a 3-focus stream checking structure in Massachusetts, a 4-focus system in Honduras and—maybe the simplest joined flood watching structure within the US—the Iowa Flood data system (IFIS). While most existing flood-watching structures are based on gigantic development conductor bowls, streak flooding has become less ideal within the WSN social sales. Marin-Perez et al. (2012) develop a 9-focus point WSN for streak flood finding during a semiarid watershed in Spain, while See et al. (2011) utilize a Zigbee-based WSN to screen void pot floods in an urban sewer framework. While most systems are still pilot-scale, these undertakings show the restriction of WSNs for scattered flood taking a look at an assortment of scales and conditions.

8.5 THE OPEN TORNADO STAGE

Open tornado gives an unquestionable and joined structure for seeing and controlling urban watersheds. Without a doubt, it's the main open-source, past what many would consider a conceivable stage that sets an advanced seeing, control and cloud relationship. The undertaking is needed to form responsibility by lowering past what many would believe workable for accomplices, pioneers, and experts. Presently, the open storm structure is joined by an assortment of reference material that hopes to enable non-specialists to send their own sensors and controllers. This living record, accessible at open-storm.org, gives instructional exercises, documentation, considered rigging, and sound examinations for arranging the board. Despite articulating center highlights, this guide details the (referencing) stray bits of sensor mastermind sending, including data that's dependably not accessible in diary articles, for instance, mounting gear, checks and affiliation frameworks. The open typhoon structure can widely be isolated into three layers: gear, cloud affiliations, and applications. The apparatus layer wires contraptions that are sent within the field, for instance, sensors for get-together disagreeable information, actuators for controlling water streams, chip, and remote transmitters. The cloud affiliations layer wires arranging utilities that get, store and procedure information, and talk with field-sent contraptions through client outlined applications. Finally, the appliance layer depicts how clients and driving models help field-sent contraptions. This three-level organizing acknowledges applications to be made at a raised level, without the need for low-level firmware programming. Together, these layers contain a versatile structure that will be straightforward and often changed by the necessities of a good gathering of clients and applications.

8.6 NOTEWORTHINESS LIMIT

The most recent IEA study assesses that half of CO_2 emissions could be decreased through energy capacity improvement. Some industry, transportation and construction, have been viewed as key regions to also improve the centrality breaking point and specialist models are given. For critical concentrated endeavors, for example, steel and iron, squash and paper and petrochemicals, increasing proficiency could be rehearsed through different turns of events. Supervisory control and data acquisition (SCADA), manufacturing execution systems (MES), and enterprise resource planning (ERP) are routinely utilized in industry to screen the creation procedure, reinforce the new development and upkeep of mechanical systems, and thus lessen criticalness use. Correspondingly to industry, the structure division harbors dazzling potential for improving centrality limit utilizing a home energy management system (HEMS): HEMS can screen and timetable home machines subject to client models and reliable power costs, improve possible segment by organizing viably and sales measure, and bolster request of building criticalness structures, especially HVAC structure operation [60]. Considered altogether more widely, related HEMS can change into a supposed strategy of-structure (SoS) to decrease the summit energy arrangements of structures in frameworks and urban spaces; such interest side connection cutoff purposes of HEMS will be referenced. For the transportation zone, methods for deduction for improving the immensity limit of standard fuel vehicles join capacity rule and tailpipe discharge control [71]; yet over the long haul, the move from internal combustion engines to electric vehicles (EV) for vehicles and light-obligation vehicles is a dependably supportive course for low-carbon mobility [72]. Previous evaluations have shown that the flood of electric vehicles relies upon the age technique of power grids [73], so developing low-carbon age in the power blend is central and as such talked about in Section 8.1. Furthermore, the association between electric vehicles and the system has a great effect on the structure and headway of force cross zones. On the one hand, EV battery charging could change the store touch of a force structure subsequently requiring power limit progress then again, the association and-play activity model of vehicle-to-filter through (V2G) could make it a potential turning put something in a sheltered spot for the redundant control of dissipating grid. For this condition, EV bunch the board structures must push toward framework wide data sharing and skimmed control to give charging approach update, singular adaptability showing up and V2G building among others. CPS would indisputably be an exceptionally colossal resource for such an update.

8.7 CLOUD AFFILIATIONS

While sensor focus can work directly by overseeing information and picking choices on a near level, getting together with cloud affiliations connects with structure scale oversight, strategy, and control of field-sent contraptions. Like a standard SCADA framework, the cloud affiliations layer underpins telemetry and the breaking point of sensor information, gives depiction cutoff focuses, and draws in remote control of contraptions—either through manual data or through electronic timetables. Regardless, not in the scarcest degree like a standard SCADA structure, the cloud affiliations layer similarly permits sensor focus fixations to visit with a wide blend of client depicted web applications—including pushed information outline contraptions, control tallies, GIS programming, outside information ingesters, organized frameworks, and constant hydrologic models. By combining consistent oversight and control with district express instruments, this structure empowers versatile framework scale control of water resources. To summarize, the cloud affiliations layer plays out the going with center cutoff centers: stores and systems remotely transmitted information, improves the board and support of field sent sensor arrange focuses, empowers mix in with a set-up of trustworthy models, control estimations, and depictions. These affiliations are condition freethinker, recommending that they can be sent on a near server or a virtual server in the cloud. In each practical sense, regardless, current open hurricane attempts are passed on striking cloud relationship, for example, Amazon Adaptable Compute Cloud (EC2) or Microsoft Azure—to guarantee that computational assets deftly scale with request. In the going with piece, we depict the basic structure, and present model applications that are gotten together with the open hurricane stage. The database fills twofold needs as both a cutoff motor for sensor information, and as a correspondence layer between field-sent sensors and web applications. The standard work of the database is to store advancing toward estimations from field-sent sensors. Sensor focus focuses report estimations direct to the database through a guaranteed web connection—utilizing a relative show that one may use to find the opportunity to webpage page pages in a program (HTTPS). The database address (URL) is appeared in the sensor place point firmware, permitting the client to make information to an endpoint subject to their own tendency. Notwithstanding overseeing sensor estimations, the database also interfaces with bidirectional correspondence among inside and cloud-based applications by overseeing contraption plan information, demand signs, and information from outside sources. Server applications talk with the sensors place by sending data to the database. These are then downloaded by the sensors on each wakeup cycle. For instance, an anticipated control application may change flood from a cutoff bowl by making a game arrangement out of valve positions to the database. At each surveying between times, the sensor spot will demand the most recent required valve position and deals the right control headway. This structure pulls in bidirectional correspondence with field-sent sensor conditions.

8.8 VITALITY OF THE BOARD STRUCTURES

As discussed in Section 8.2, another zone where CPS advancements are important is making hugeness probability, as an example in land the heads or the skilled utilization of noteworthiness structure. Building the executives structures, within the structure part, CPS will change practices when molded into building the board frameworks (BMS). The Brick outline may be a specialist CPS-based BMS application around there; its essential issue is to deal with the sound data of sensors, frameworks and building structures in the existing structure, the pros structures (BMS) through class chain of importance (mark sets) and relationship sets. The Brick chart is found during a favored position depiction structure information model that watches out for information as triples: subject, predicate, and object. Specifically, Brick progress applications enable us to coordinate stuck damper attestation by watching viable wind stream sensor respects. BOnSAI, a platform for AI, is another general CPS application performing at the structure level. Almost like Brick design, BOnSAI targets sharp structures, yet from the point of view of wrapping data (an inevitable, re-try, setting careful enrolling condition through inserted IoT framework in structures). The classes in BOnSAI are built into several standard contemplations: gadget, association and setting. Gadget class delineates the contraptions and mechanical social affairs as slightly of the physical substances in structures, for instance, air system, lighting, sensor and actuator; association class plots the functionalities of gadgets as activities, with each activity having its own information, yield, precondition and impact; setting class diagrams the bits of varied endeavors in various conditions. Furthermore, BOnSAI appeared to engage the built control of SmartPlugs on faculty grounds by unraveling the sensor parameters at different areas.

8.9 CASH RELATED AND NORMAL EFFECT ASSESSMENT

Beginnings of energy markets, business models, and use structures are being flipped around and new suppliers, for instance, stage progress from other section are legitimately entering the market. Notwithstanding this difficulties of progress, new advances are influencing inside business culture, frameworks and therefore the general relationship of the centrality relationship. For instance, generally interests in bleeding edge hierarchy and programming increased by 30% once a year in 2017. The cash-related objective behind these undertakings is clear: the value spare records potential of CPS and its subsystems is moved to be within the space of US$80 billion, within the level of 2016 and 2040. A big fragment of the abatement potential may be a result of decreased assignments and maintenance costs, probability refreshes correspondingly as diminished individual events and yielded lifespans [28]. Consistent with the IEA, the four zones are the crucial providers: Cyber-dangers to imperativeness and mechanical security. The epic positive conditions discussed in Section 8.1, after a brief time, accompany clear detriments: while CPS will improve the endeavors of basically principal parts, for instance, centrality and industry, it'll all the while make them fragile against cyberattacks and subsequently robotized dependent. Hostile elements could really impact national economies by upsetting the intentionally fundamental agilely of vitality at whatever point over the immensity respect chain. CPS are along these lines not just the empowering effect of plentifulness refreshes, enlarged quality and raised security, yet simultaneously increment the deferred results of a possible threat. For instance, the industrial control system (ICS) might be harmed, which might incite loss of control of central foundation and henceforth could cause dissatisfaction with an important resource.

8.10 CONCLUSION

The envisioned complexity in significance structures towards presence of mind has three principal estimations materials required to thrive: (i) low-carbon power plans, (ii) essentialness attainability improvements, and (iii) centrality gathering task. This chapter gives clear models exhibiting that CPS developments are going to be fundamental for the gainful utilization of everything of the three estimations. The first occurrences of the usage of AI into CPS have demonstrated that this synergetic blend can have disturbing great conditions and can lead undeniably to budgetary and run-of-the-mill gains, but may affect social risks that are unforeseeable when created. Critical appraisals were wont to depict the upsides of CPS and its sub-structures for the cash-related and normal sensibility of the quality pathways to the low-carbon significance systems named starting at now. On the off chance that there should rise an occasion of renewables, CPS will affect the blend of the unpredictable essentialness source in existing centrality structure, as an example through PC helped unfathomable resources obvious evidence and satisfactory force source induces improved by AI movement. By goals of imperativeness efficiency, CPS are having an impression that's in every way that basically matters bound to develop a very fundamental level over the approaching years; this effect is exemplified using advanced building the board structures, the development of server creates and invigorated intrigue side affiliation. Open tornado's enhancement of wide configurability, steady response, and automatic control make it a perfect choice for water structure heads and standard experts the indistinguishable. While many open gear stages exist, open typhoon is the standard open-source, thoroughly stage that concretes perceiving, control and energy monitoring systems (EMS). EMS utilize dynamic structures that empower correspondence between interfacing pieces within the system. A schematic of EMS is shown and it will, generally speaking, be seen that the growing affirmation of the advanced metering infrastructure (AMI), alongside data-driven decision help, could allow billions of machines to be related to demand response. Specifically, with reference to HEMS, the EMS could get respect signals from structure bosses or aggregators, which might be treated as judges between the cross-region head and end customers. The HEMS would make decisions on how different contraptions should be made game arrangements for mentioning to assist utility without ambushing essential limit.

Continuously, the control dissipated takes care of the relationship of water resources the board. Almost giving a mechanical strategy, open whirlwind looks out for this force reality nuts and bolts, which will be standard in water resources applications, for instance, field-power, low power advancement and structure scale coordination. The open tropical storm experience has exhibited leads to build up the wants of existing storm water structures: both by growing the spatiotemporal goals of estimations, and by sufficiently improving water quality through decided control. Regardless, open whirlwind is not just a phase—it's in like way a game plan of pros, partners and pioneers who are supported seeing ceaselessly stunning water structures. to assist the spread and development of wonderful water systems, we are becoming by report at open-storm.org so as to share standards, reference materials, structures, use cases, evaluation estimations, and other strong resources. We invite customers to research this undertaking by offering their experiences of arranging, abandoning, and keeping water structures. The chapter is a brief look at whether mechanized degrees of progress, maintained by man-affected speculation, can push decarbonization direction onto courses that would avoid harming impacts from typical change and even trade the strategy.

BIBLIOGRAPHY

[1]   Aazam, M., and Huh, E.-N. (2016). Fog computing: The cloud-IoT/IoE middleware, *IEEE Potentials*, 35, 3, May–June 2016, 40–44

[2]   Kortuem, G., Kawsar, F., Fitton D., and Sundramoorthi, V. Smart objects and building blocks of internet of things. *IEEE Internet Comput. J.*, 14, 1, 44–51.

[3]   Aazam, M., Zeadally, S., and Harras, K. A. (2018). Fog computing architecture, evaluation, and future research directions. *IEEE Communications Magazine*, 56, 46–52.

[4]   Abomhara, M., and Koien, G. M. (2014). Security and privacy in the internet of things: Current status and open issues. *International Conference on Privacy and Security in mobile Systems (PRISMS)*, Aalborg, Demark, 1–8.

[5]   Abujubbeh, M., Al-Turjman, F., and Fahrioglu, M. (2019). Software-defined wireless sensor networks in smart grids: An overview. *Sustainable Cities and Society*, 51, 101754.

[6]   Adelman, D., and Mersereau, A. J. (2008). Relaxations of weakly coupled stochastic dynamic programs. *Operations Research*, 56, 712–727.

[7]   Akeela, R., and Elziq, Y. (2017). Design and verification of IEEE 802.11 ah for IoT and M2M applications. In *IEEE International Conference on Pervasive Computing and Communications Workshops (PerCom Workshops)*, Kona, USA: IEEE,USA 491–496.

[8]   Seth, B., and Dalal S. (2018) Analytical assessment of security mechanisms of cloud environment. In Saeed, K., Chaki, N., Pati, B., Bakshi, S., and Mohapatra, D. (eds) *Progress in Advanced Computing and Intelligent Engineering*, Advances in Intelligent Systems and Computing, vol. 563. Singapore: Springer. https://doi.org/10.1007/978-981-10-6872-0_20, 211-220.

[9]   Al-Doghman, F., Chaczko, Z., Ajayan, A. R., and Klempous, R. (2016). A review on fog computing technology. *2016 IEEE International Conference on Systems, Man, and Cybernetics (SMC)*, IEEE, 001525–001530.

[10] Al-Fuqaha, A., Guizani, M., Mohammadi, M., Aledhari, M., and Ayyash, M. (2015). Internet of things: A survey on enabling technologies, protocols, and applications. *IEEE Communications Surveys and Tutorials,* 17, 2347–2376.

[11] Alrawais, A., Alhothaily, A., Hu, C., and Cheng, X. (2017). Fog computing for the internet of things: Security and privacy issues. *IEEE Internet Computing*, 21, 34–42.

[12] Al-Sarawi, S., Anbar, M., Alieyan, K., and Alzubaidi, M. (2017). Internet of things (IoT) communication protocols. In *8th International Conference on Information Technology (ICIT)*, IEEE, 685–690.

[13] Altieri, A., Piantanida, P., Vega, L. R., and Galarza, C. G. (2015). On fundamental trade-offs of device-to-device communications in large wireless networks. *IEEE Transactions on Wireless Communications*, 14, 4958–4971.

[14] Seth, B., Dalal, S., and Kumar, R. (2019) Hybrid homomorphic encryption scheme for secure cloud data storage. In Kumar, R., and Wiil, U. (eds) *Recent Advances in Computational Intelligence*, Studies in Computational Intelligence, vol. 823. Cham: Springer. https://doi.org/10.1007/978-3-030-12500-4_5.

[15] Al-Turjman, F. (2019). Cognitive routing protocol for disaster-inspired internet of things. *Future Generation Computer Systems*, 92, 1103–1115.

[16] Al-Turjman, F., and Malekloo, A. (2019). Smart parking in IoT-enabled cities: A survey. *Sustainable Cities and Society*, 49, 2019, 69–78.

[17] Arshad, R., Zahoor, S., Shah, M. A., Wahid, A., and Yu, H. (2017). Green IoT: An investigation on energy saving practices for 2020 and beyond. *IEEE Access: Practical Innovations, Open Solutions*, 5, 15667–15681.

[18] Asheralieva, A., and Niyato, D. (2019). Game theory and Lyapunov optimization for cloudbased content delivery networks with device-to-device and UAV-enabled caching. *IEEE Transactions on Vehicular Technology*, 68, 10094–10110.

[19] Atlam, H. F., J. Walters, R., and Wills, G. B. (2018). Fog computing and the internet of things: A review. *Big Data and Cognitive Computing*, 2, 10.

[20] Atzori, L., Iera, A., and Morabito, G. (2010). The internet of things: A survey. *Computer Networks*, 54, 2787–2805.

[21] Aust, S., Prasad, R. V., and Niemegeers, I. G. (2012). IEEE 802.11 ah: Advantages in standards and further challenges for sub 1 GHz Wi-Fi. In *IEEE International Conference on Communications (ICC)*, Egypt, 6885–6889.

[22] Balevi, E., and Gitlin, R. D. (2018). Synergies between Cloud-fag-thing and brain-spinal cord-nerve networks. Information Theory and Applications Workshop (ITA), IEEE, 1–9.

[23] Bangui, H., Rakrak, S., Raghay, S., and Buhnova, B. (2018). Moving towards smart cities: A selection of middleware for fog-to-Cloud services. *Applied Sciences*, 8, 2220.

[24] Bastug, E., Bennis, M., and Debbah, M. (2014). Living on the edge: The role of pro-active caching in 5G wireless networks. *IEEE Communications Magazine*, 52, 82–89.

[25] Bastug, E., Bennis, M., and Debbah, M. (2015). A transfer learning approach for cache enabled wireless networks. In *13th International Symposium on Modeling and Optimization in Mobile, Ad Hoc, and Wireless Networks (WiOpt)*, Kabul (Pakistan), 161–166.

[26] Bilal, K., Khalid, O., Erbad, A., and Khan, S. U. (2018). Potentials, trends, and prospects in edge technologies: Fog, cloudlet, mobile edge, and micro data centers. *Computer Networks*, 130, 94–120.

[27] Seth, B., Dalal, S., and Kumar, R. (2019). Securing bioinformatics cloud for big data: Budding buzzword or a glance of the future. In Kumar, R., and Wiil, U. (eds) *Recent Advances in Computational Intelligence*, Studies in Computational Intelligence, vol, 823. Cham: Springer. https://doi.org/10.1007/978-3-030-12500-4_8, 268-275.

[28] Bonomi, F., Milito, R., Natarajan, P., and Zhu, J. (2014). Fog computing: A platform for internet of things and analytics. In *Big Data and Internet of Things: A Roadmap for Smart Environments*. Berlin: Springer, 169–186.

[29] Brogi, A., Forti, S., and Ibrahim, A. (2017). How to best deploy your fog applications, probably. *IEEE 1st International Conference on Fog and Edge Computing (ICFEC),* Koria, 105–114.

[30] Calheiros, R. N., Ranjan, R., Beloglazov, A., De Rose, C. A., and Buyya, R. (2011). CloudSim: A toolkit for modeling and simulation of cloud computing environments and evaluation of resource provisioning algorithms. *Software: Practice and Experience,* 41, 23–50.

[31] Cao, Y., Chen, S., Hou, P., and Brown, D. (2015). FAST: A fog computing assisted distributed analytics system to monitor fall for stroke mitigation. IEEE International Conference on Networking, Architecture and Storage (NAS), IEEE, 2–11.

[32] Ceipidor, U. B., Medaglia, C., Marino, A., Morena, M., Sposato, S., Moroni, A., et al. (2013). Mobile ticketing with NFC management for transport companies: Problems and solutions. 5th International Workshop on Near Field Communication (NFC), IEEE, 1–6.

[33] Ceipidor, U. B., Medaglia, C., Volpi, V., Moroni, A., Sposato, S., Carboni, M., et al. (2013). NFC technology applied to touristic-cultural field: A case study on an Italian museum. 5th International Workshop on Near Field Communication (NFC), IEEE, 1–6.

[34] Chang, Z., Lei, L., Zhou, Z., Mao, S., and Ristaniemi, T. (2018). Learn to cache: Machine learning for network edge caching in the big data era. *IEEE Wireless Communications,* 25, 28–35.

[35] Chen, M., Mozaffari, M., Saad, W., Yin, C., Debbah, M., and Hong, C. S. (2017). Caching in the sky: Proactive deployment of cache-enabled unmanned aerial vehicles for optimized quality-of-experience. *IEEE Journal on Selected Areas in Communications,* 35, 1046–1061.

[36] Chiang, M., Ha, S., Chih-Lin, I., Risso, F., and Zhang, T. (2017). Clarifying fog computing and networking: 10 questions and answers. *IEEE Communications Magazine,* 55, 18–20.

9 Social Distance Monitoring in Smart Cities using IoT

J. S. Kalra, Rajesh Pant, Shipra Gupta,
and Vijay Kumar
Corresponding Author Dr. Vijay Kumar

CONTENTS

9.1   INTRODUCTION

Covid-19 has challenged humankind by infecting a large population of world, and as of April 3, 2020 more than half of the world's population was under lockdown [1]. These times always cause people to contemplate present obstacles and search for solutions. Present technology and infrastructure need to be modified and excelled to contain this contagious virus. Covid-19 has had unforeseen effects on the globe. A study by a global tech market advisory firm forecasted that the manufacturing revenue of world will be cut by US$15 trillion due to non-working of supply chains for 2020 [1].

The absence of direct medical assistance in Covid-19 means infected people must be stopped from coming into contact with healthy people following social distancing. This strategy involves country-wide lockdowns, quarantine, travel bans, hotspot containment, reducing the numbers of customers and managers in businesses, but these measures cannot be followed in long run as they will strip the country's economy and could lead to a much more disastrous situation than that posed by the virus alone. Successful social distancing was observed in Singapore which has effectively prevented the community spread during SARS-CoV-2 [3]. But the response of China to Covid-19 was far more extreme, including closure of workplaces, schools, roads, travel ban, compulsory home quarantine for uninfected people, separate quarantine to infected people, and wide-scale electronic surveillance [4, 5]. It is difficult to give

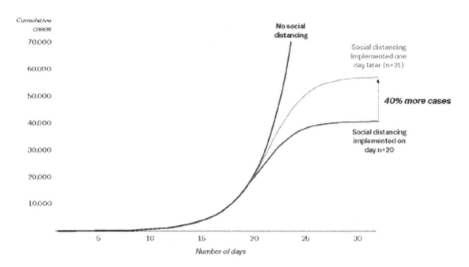

FIGURE 9.1   Effect of social distancing

a speedy solution in such critical times, and numerous agencies are working with government to contain the spread of virus. A comprehensive self-alert intelligence system is required to ensure the wellbeing of people with the purpose of containing the infection and reducing the mortality rate.

Even when mass movement was curbed and strict rules were issued to follow social distancing, the numbers of cases were increasing, which clearly shows that we are not well prepared for averting such pandemics. During such a situation, quarantining infected people, keeping a track of and monitoring those who have come into contact with them, minimizing the movement, and increasing awareness on following social distance guidelines are the measures which should be coupled with technology to make a robust system to fend off such diseases. In a few initiatives worldwide countries used drones to sprinkle sanitizer in infected areas and monitor the movement of people. Body wearable devices [6] were used to keep a check on the population and mobile applications monitored mass gatherings. Contact tracing applications are being developed which will trigger an alarm if someone comes in contact with an infected person [7, 8].

Many countries are developing "smart" cities, including USA, India, China, and Japan, which could give an edge over such pandemics. For example, Amsterdam's model of smart city 3.0 [9] motivates citizens to participate in creating smart solutions which include intelligent transportation system (ITS), smart energy, research, data sharing in advancement of technology and infrastructure-related solutions. Such cities with a wide range of data driven services, along with sensors and ITS, will not only help in mitigating post pandemic effects but will also provide a robust solution for early detection of outbreaks. Countries like India are working towards smart cities by providing a technological solution to problems like waste management, traffic issues, Wi-Fi networks, and crime control by using AI based CCTV cameras. With the use of the latest techniques in smart cities and the present pandemic situation the technology should be harnessed for the implementation and maintenance of social distancing within a smart city.

9.2 METHODOLOGY

Various smart technologies can be combined together for social distance monitoring and implementation within a smart city. One of those technologies could be a Mandatory Smart Health Band (MSHB), which could be made compulsory within the administrative boundaries of a smart city. Every citizen of that smart city will have his/her own unique individual MSHB. Mandatory wearing of this MSHB could be enforced by law, and punishments could be set for not wearing MSHB in public places. Secondly, sensors installed in all public places, such as malls, religious places, parks, markets, could sense the number of MSHB present in a particular area. The person obtaining a new MSHB would have to register his/her contact details which would be linked to the particular MSHB.

9.3 APPLICATION OF MSHB

It will eliminate the need for a person to carrying various IDs, as all their identifying details will be linked to the MSHB. Whenever he/she is required to produce an identification document a QR code could be generated in the on MSHB screen and scanned on an authorized device which is connected to the data server. So it eliminates the need of carrying a number of documents in your pocket.

If in a particular location there are more than the allowed number of persons, a scanner installed at every public gathering place will send a warning message to all the MSHB owners there and if they do not move from there, it will trigger an alarm message to the police control room for the necessary action. Thus it can prevent unauthorized social gathering at public places.

If a person is home quarantined, that detail will be uploaded in his/her central data, and will be immediately reflected wherever his/her QR code is scanned. Scanning of the MSHB QR code will be a common practice as a vehicle's document checking is done. The scanning device will be an authorized person's phone. A person will be as careful to carry their MSHB as he/she fears driving without a valid driving license.

If a person is home quarantined and he/she comes in contact with another citizen, it will trigger an alarm to that person to be wary of surrounding persons. This will prevent adding one more Covid-19 positive case.

It can also help in contact tracing of positive cases, using GPS and Bluetooth technology and automatically sending them a SMS warning them of the situation and also giving the data to the authority for the necessary action to be taken.

It is always said that if god closes one door, many others are opened. Although everyone is aware of the coming recession and unemployment and even the complete shutdown of many business houses, there is an opportunity for a new business vertical, starting from manufacturing of MSHB, infrastructure development, deployment, maintenance of infrastructure as well as maintenance of MSHB. Thus it could also be seen as a new horizon of business and a turning point for new start-ups.

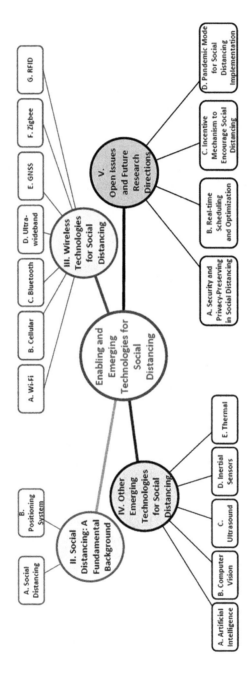

FIGURE 9.2 Emerging technologies for implementation and monitoring of social distancing [8]

FIGURE 9.3 MSHB displaying the QR code

FIGURE 9.4 Public gathering can be controlled using MSHB

FIGURE 9.5 Documents scanning will be replaced by MSHB scanning

FIGURE 9.6   Contact tracing with the help of MSHB

9.4   WORKING AND FEATURES

The MSHB will have inbuilt temperature sensor, pulse monitor, GPS, Bluetooth, and Wi-Fi and it will be connected to a server. All the recorded data regarding health, contact with any Covid-19 patient, contact tracing, violation of social distancing rules, etc., will be sent to the server, from where on the basis of a machine learning algorithm defaulters' details will be sent to their nearest police station, from where the required necessary legal action could be taken.

If a person is in home quarantine, his/her record can also be monitored. If he leaves the house a machine learning algorithm can alert the nearest police station. This feature can also be used in a mobile application, but people can leave their phone at home. In the case of MSHB, it is mandatory and can be randomly checked by authorities anywhere, as stated above.

If a person is found to be Covid-19 positive his/her recent close contacts can be traced and contacted through the database maintained in the server. The close contacts can be triggered with a SMS through a machine learning algorithm, to be in self-quarantine for next 14 days. The contacted persons will automatically be included in the list of home-quarantine persons, and the particular algorithm for home-quarantine people will run for them. Their condition can also be monitored by using the temperature reading from MSHB.

To detect the violation of social distancing sensors can be installed at places like malls, high streets, parks, bus stops, etc., which will sense the presence of number of MSHB in a particular location and will trigger an alert to the nearest authorities to for an instant response.

At present it is a big question for authorities whether to test a particular person or not, but after the mandate of MSHB it will automatically prompt the authorities to test a particular person on the basis of his/her contact history along with the temperature monitoring on the server.

It will also alert the user with vibration if there are chances of coming in near contact (less than 10 meters) with a Covid-19 positive or risky patient or a containment zone, using GPS and Bluetooth technology.

The MSHB will generate a QR code on its screen which can be scanned at any check point before entering premises like mall, university, stadium, office, etc. Scanning will be connected to a server and if it shows green that means the person

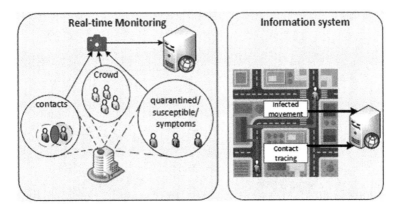

FIGURE 9.7 Real-time monitoring [8]

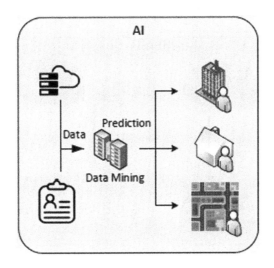

FIGURE 9.8 Use of AI, datamining and machine learning algorithms [8]

is not at any risk and can be granted entry. On scanning it will give all details of the person's identity, which will make sure person is wearing his/her own MSHB not someone else's.

The distribution of MSHB will be done by the smart city authority. Its manufacturing maintenance work can be given to young start-ups to boosting the start-up culture within the country.

9.5 CHALLENGES

There cannot any bigger challenge than Covid-19 itself, but still there will be many small challenges faced for implementation of this technology on 100 percent of the population and area of a smart city.

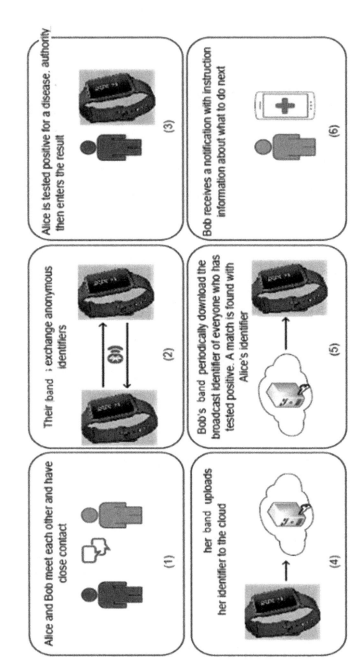

FIGURE 9.9 Use of the band

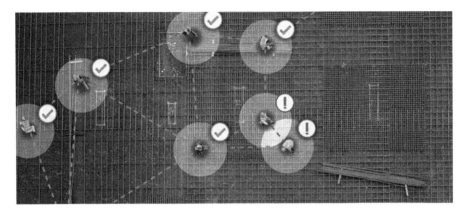

FIGURE 9.10 Alert on close proximity of a Covid-19 positive or risky person [8]

Infrastructure Challenge: Setting up the required infrastructure will be the first challenge to be faced. Infrastructure challenges such as setting up a server connected to whole city, installation of sensors for sensing the number of MSHB at places of possible public gathering, setting up of check points at the entry of any public premises, providing of scanning machines to police persons.

Training Challenge: Training of the required manpower for running and maintenance of the complete system will also be one of the minor challenges.

Awareness Challenge: Awaking people for the MSHB will also be one of the initial challenge as people will not choose to purchase this band and if purchased they will not wear it when leaving home. This can only be achieved by making it mandatory in law.

Political: Implementation of a new law involving the whole population has never been an easy task for any government in any country. There are always objections from opposition parties, for some reason or other. Thus it will also be a challenge to overcome the political pressure from opposition political parties.

Acceptance: Once things have been done at the government end, and time comes for implementation, citizens are not ready to accept it. Either the users are ignorant or they are careless, so it will be a big challenge to reach each and every citizen in time so that implementation of MSHB can be done quickly for the safety of citizens.

Maintenance: After successful implementation there will be the further challenge of maintenance of these many MSHBs which will require setting up of service centers, training of employees, arranging of spares, etc.

Future Scope: Though the designing of the MSHB has kept minute details in mind and looked at each and every challenge, but still the scope of improvement

FIGURE 9.11   Future of MSHB

should never be neglected. For instance this can presently be implemented in smart cities but in future it could be extended to every city. It may be integrated in holographic technology, virtual reality, and there may even be in body implantation in future.

9.6   CONCLUSION

It is well known that social distance is mandatory to prevent the spread of coronavirus Covid-19. The virus can attack the human beings when the distances between people are less than 1 meter. To implement social distance between people, the governments of all the countries have applied lockdown in their cities and states. Many people do not follow the rules of government and due to this the number of infected cases continues to rise. To monitor social distance the internet of things in smart cities will be very useful. It is concluded that the problem of maintaining and monitoring social distances may be reduced by the application of the latest communication and AI technologies. Technology has always been a savior of mankind, and we will win this fight against the deadly Covid-19 by successful and novel application of technology.

BIBLIOGRAPHY

[1]   Euro News, ABI research. https://iotbusinessnews.com/2020/03/17/07945-coronavirus-outbreak-reveals-the-weakest-links-in-the-supply-/chain-the-suppliers-supplier/ (accessed 11 April 2020).

[2]   Koo, J. R, Cook, A. R., Park, M., et al. Interventions to mitigate early spread of COVID-19 in Singapore: a modelling study. *Lancet Infect Diseases*, 2020; published online March 23. https://doi.org/10.1016/S1473-3099(20)30162-6.

[3] Kupferschmidt, K, and Cohen, J., China's aggressive measures have slowed the coronavirus. They may not work in other countries. 2 March 2020. www.sciencemag.org/news/2020/03/china-s-aggressivemeasures-have-slowed-coronavirus-they-may-not-work-other-countries (accessed 9 March 2020).

[4] Kupferschmidt, K., and Cohen, J., Can China's COVID-19 strategy work elsewhere? 6 March 2020. https://science.sciencemag.org/content /367/6482/1061?r ss%253D1= (accessed 11 March 2020).

[5] SPHCC employs IoT tech and wearable sensors to monitor COVID-19 patients. www.mobihealthnews.com/news /asia-pacific/ sphcc-employs-iot-tech-and-wearable-sensors-monitor-covid-19-patients (accessed 11 April 2020).

[6] Apple, Google bring Covid-19 contact-tracing to 3 billion people. www.bloomberg.com/news/articles/2020-04-10/apple-google-bring-covid-19-contact-tracing-to-3-billion-people (accessed 11 April 2020).

[7] Coronavirus: Singapore develops smartphone app for efficient contact tracing. www.straitstimes.com/singapore/coronavirus-singapore-develops-smartphone-app-for-efficient-contact-tracing (accessed 11 April 2020).

[8] Amsterdam Smart City 3.0. https://smartcityhub.com/governance-economy/amsterdam-better-than-smart/ (accessed 12 April 2020).

[9] Nguyen, C. T., Saputra, Y. M., Van Huynh, N., Nguyen, N. T., Khoa, T. V., Tuan, B. M., … and Chatzinotas, S., Enabling and emerging technologies for social distancing: A comprehensive survey. 2020. arXiv preprint arXiv:2005.02816.

[10] Kokalki, S., Mali, A., Mundada, P., and Sontakke, R., Smart health band using IoT. *IEEE ICPCSI*, 2017, pp. 1683–1687. 10.1109/ICPCSI.2017.8392000.

[11] Friedman, C. P., A "fundamental theorem" of biomedical informatics. *Journal of the American Medical Information Association*, 16(2), 2009, p. 169.

10 Sustainable Solution for Load Forecasted EHV Charging Station in Smart Cities

Ajay Kumar, Parveen Poon Terang, and Vikram Bali

CONTENTS

10.1 INTRODUCTION

In earlier days, the unidirectional or one-way flow of energy occurred between the main power generation system to consumers through transmission lines [1]. As per the total requirements, in terms of load, the required electrical power was obtained by manipulating the operation of the power plants. Due to the monopoly of the government over all aspects of the energy sector viz., generation, transmission, distribution and trading, there were no changes. Nowadays, with the latest tools and technologies, new inventions and innovations, the distributed network of power called the grid come into the picture. It is the largest human-made new bridge of the twenty-first century, still undergoing transformation and rejuvenation, making it even more irrepressible, stable and smart. Efficient delivery of economical, secure and sustainable power has come about due to the intelligent integration of generation, transmission, and distribution in a smart grid [2]. It provides an intelligent power system, which comprises an advanced metering infrastructure with the help of various smart and automatic applications. It makes the process of demand response and distribution management system smart. It uses an intelligent advanced energy storage system by integrating electric vehicles and advanced electricity marketing [1].

The commencement of the twenty-first century saw the concept of the smart city come into existence. This was made possible only by realization of consumer-friendly information and technologies for communication developed by major industries. It relates to the development of cities and makes them smart by using new tools and techniques. By using the power of machine learning and IoT, cities are now progressive, becoming smarter and resource efficient. Smart cities provide quality of life and promote innovations on both the social and technical front, making optimum use of existing infrastructures.

Sustainable green solutions are incorporated leading to a shift in focus towards consumer participation and new forms of governance. They are smart with an ability to make intelligent decisions at the strategic level and they consist of large individual projects and long-term implementations to do this. When cities are considered as one complete system, the goal of becoming smart is achievable. The smart cities then take on the challenges of resource scarcity and climate changes in a forceful united manner. This cascades into the need to sustain quality of life and economic competitiveness for the ever-growing demand of the urban population.

10.2   GREEN SMART CITIES

Integration of renewable energy sources enables the power generation system to be decentralized resulting in a multi-directional power flow, i.e. the power flow between consumers and power generation system is in various directions through multiple transmission lines. The consumers can also give inputs in generating power and actively participate in electric power organization. They can modify their electricity purchasing pattern and as per their behavior, they can be informed and have choices, agreement and disagreement. These choices depict the change in technologies and markets. There is an increase in competitiveness, answerability, and responsibility due to large-scale private participation of the government-centric energy sector. Internet of Things (IoT) enables the daily usage entities to be interconnected with the help of network based infrastructure [11].It is used as a self-organizing map to connect various devices of everyday usage..

It connects the sensors, electronic devices, and codes of various dimensions to the wireless networks with the help of some interface. This technology provides a protocol for communication between human and machine or machine to machine through various schemes.

There are various aspects of IoT, the most attractive of which are, intelligence, and network/internet connect [7][8][9][10]. It enables the data-gathering or data-acquisition system to be more interactive and secure by using communication which is bilateral, handling and response control. There are so many examples where IoT comes into the picture and plays a vital role in making the system intelligent;

example, when a driver drives the car in an unskilled manner, the vehicle will buzz an alarm; a forgetful passenger will be prompted by the suitcase about neglecting something; communication from clothes to washer about the color and desired temperature and so on. The same will influence the choices of loads or peak loads in grid management and electric power use in the form of smart grids. But there are some challenges, which arise while trying to regularize the load, along with higher penetration and increased distribution of non-conventional energy sources in terms of equalizing expenditure and production. The use of storage devices, which have intelligent control, in distribution grid will improve the performance of the system and also improve the process.

There is a need for distributed energy storage in order to handle a bulk energy storage system in the utility grid. Due to this, integration of renewable energy sources is required. As per the United Nations, 67% of the world's population will no longer be living in rural areas by 2050. This will create a demand in the transportation sector, thereby cascading into a sharp increase in usage of plug-in hybrid electric vehicles (PHEVs) and electric vehicles (EVs). The selection of non-conventional energy sources as sustainable fuel for these will be accelerated with the growing demands [19][20].

10.3 ELECTRIC VEHICLES

In EVs, the internal combustion engine is replaced by electric motors, which make the advent of EVs different from fuel-based conventional vehicles. There are three parameters or sources for the energy used to drive motors, which are power generation plants, on-board generators, and on-board energy storage system. An on-board energy system is the most valuable source for electric vehicles [4]. In the present scenario, more stress is on the plug-in feature of EVs by taking the inputs from automobile industries and academia.

As per the report by the International Energy Agency, Global EV Outlook 2019, "Scaling up the transition to electric mobility," the deployment of electric vehicles is a growing trend.

With an increase of approximately 63%, the issues of charging and smart control increase. The emissions that affect the sustainability of EVs is also an aspect that needs to be considered given the scenario of the rapid urban transition.

Minnesota Pollution Control Agency looks into the vehicle emissions by fuel type. Categorically, electric vehicles can be classified as battery electric vehicles (BEVs), i.e. main feature is battery enabled and plug-in hybrid electric vehicles (PHEVs) in which the main feature is plug-in enabled.

10.3.1 BATTERY ELECTRIC VEHICLES

BEVs are constituted of three major components: (1) electric motor, (2) battery, and (3) controller. All three components have different functionalities. The electric motor generates the electric propulsion that produces thrust to push the car forward. The electric battery is the source of energy for electric motor. This battery is rechargeable. A two-quadrant controller is used to manage the power supply for the electric motor due to which the vehicle can move in a forward or backward direction. This controller can be a four-quadrant controller at a particular time instant when regenerative braking is required. One of the most important components is the inverter and it plays an important role in using AC motors equipped by EVs because AC motors are less expensive and simple to use.

10.3.2 PLUG-IN HYBRID ELECTRIC VEHICLES

Hybrid electric vehicles are divided into three categories as per the different types of power trains. One of the categories is parallel hybrid, which is most commonly used and it generates the propulsion with the help of both an electric motor and additional ICE (internal combustion engine). This parallel propulsion makes the hybrid electric vehicle more powerful because it enables the vehicle to operate in two modes. One of the modes is charge-depleting mode and another one is charge-sustaining mode. When the vehicle is operated in charge-depleting mode, the vehicle can have the energy stored in on-board battery packs and help the vehicle to continue driving because vehicle can switch to charge-sustaining mode if the battery state of charge is depleted up to a threshold value. Therefore combination of these two modes improves the efficiency of vehicle and utilizes the ICE system for propulsion [14][15]. The SOC level of the battery is then monitored and maintained at a pre-defined level by the controller.

The second category is series hybrid, which is commonly used in fuel-based vehicles, and the third category is power-split hybrid.

10.3.3 BATTERY CHARGERS FOR EVs

The battery chargers for EVs play an important role in the working efficiency [3]. The Li-ion battery is one of the most usable batteries to charge parallel hybrid vehicles because the Li-ion battery is the good energy source. Its long life span and high energy density makes it economical and a preferred choice. It has less environmental influence and therefore it is environment friendly.

One of the challenges is to design the chargers for these Li-ion batteries because it requires intricate control of voltage and smooth current output. Otherwise, a large voltage and any current spikes may damage the battery [5][6]. Therefore, there is a need to design an advanced battery charger that can be useful in handling high voltage to accelerate the production of EVs. These charges are currently classified into two categories, i.e. on board or off board, one-directional (charger to the battery) or bidirectional (power can be drawn from and also injected back to the grid) [12].

On-board chargers are less effective and have restricted usage due to their cost effectiveness, weight, and occupied space. In contrast, off-board chargers are smaller in size and cost effective. So off-board chargers are a good option for EVs in terms of space and cost. Further differentiating features of battery chargers are their power flow directions. They can be unidirectional or bidirectional. A unidirectional charger requires less hardware and is simple to connect compared to bidirectional chargers. In the current scenario, most of EVs use single-phase unidirectional chargers for level

TABLE 10.1
Different EV and the parameters affecting their choice

Vehicle type	Electric vehicle (EV)	Gasoline-powered (internal combustion)	Plug-in-hybrid (PHEV)	Hybrid (HV)
Energy source	Electric only	Gasoline only	Main: Electric Sub: Gasoline	Main: Electric Sub: Gasoline
Propulsion mechanism	Motors	Engine	Combination of motor + engine	
CO_2 emissions	None	Yes	Yes	Yes
Fuel facility locations	Charging stations	Gas stations	Gas stations, chargers	Gas stations
Tax liability	Low	High	Low	Low
Cruising distance	Short	Long	Long	Long

Source: www.rohm.com/blog/-/blog/id/7172825

1 and level 2 charging. Therefore, there is a need to design bidirectional chargers by using the vehicle-to-grid (V2G) technique. Thr V2G technique ideates that an electric vehicle can inject power back to the grid [16][17].

A comparative table is given in Table 10.1. It shows the vehicle type and the parameters, which affect the choice of vehicle.

10.4 IMPLICATIONS

The climate changes have increased the awareness and concern of the environment and the hunt for alternatives to fossil fuels is primary now. This makes all renewable energy resources (RESs) that have low carbon footprint, are environment friendly, and sustainable alternatives stand out. The two most prominent sources of renewable energy are solar and wind energy. An overview of these RESs is given in the following section.

10.4.1 SOLAR ENERGY

Energy from the sun can be harnessed and converted into electrical energy by photovoltaic panels (PV) or concentrating solar power (CSP) plants.

PV systems use solid state semi-conductors to directly convert energy from the sun into electrical energy, while mirrors are used to concentrate the sun's rays at a particular point thereby increasing the temperature, which in turns drive turbines or engines (steam) to generate electricity in CSP plants. With improved techniques, the popularity of PV solar plants has increased. The basic element of the PV system is the Solar cell. The efficiency of this solar cell varies negligibly even at varying temperatures making it popular. In a practical situation, to combat fluctuations in temperature, various insulation techniques are used resulting in a small variation of efficiency with change in ambient temperature. This is the primary reason why in most cases efficiency of a solar cell is assumed to be constant.

To reduce the fuel emissions, solar PV systems have been incorporated into the transportation sector, giving rise to electric vehicles (EVs). As per reports of National Renewable Energy Laboratory (NREL), there will be decreased adverse effects on generation of electricity for mass production of EVs. The United States of America has incorporated mass production of EVs, which will amount to almost 50% of the total vehicles on the road by the year 2050. This mass increase however, will only give an 8% increase in power generation and a 4% increase in the capacity of generation. The plus would be a tremendous decrease in the carbon footprint from emissions of conventional vehicles as well as reduction in fuel usage for transportation sector.

10.4.2 WIND ENERGY

A popular choice of a RES is wind energy. As a renewable energy source, it has become popular as it complements other RESs to make a hybrid system. This takes care of the intermittency of RESs, which can be said to be one of the biggest shortcomings of RESs. According to the Global Wind Energy Council, the capacity of wind power throughout the world reached a capacity of 318,137 MW in 2013. In the past five years, there has been a tremendous increase of almost 200,000MW. The goal set by The European Wind Energy Association is meeting 23% of the power demand with wind energy by 2030. The choice of the location of wind farms is instrumental for an efficient wind energy system [13]. Further to choice of location, the quality and quantity of wind energy is dependent on wind speed [18]. Evaluation of wind speed is usually done by using the testing equipment at a height much lower than the actual height (e.g.,15 m).

10.5 CHALLENGES

Adopting EVs on the grid:

1. With increased load due to charging of EVs, there is a direct impact on the working temperature of the transformers. This increase in heating takes a toll on the lifespan of the transformers incurring extra expenditure. Choice of RERs should also be made from those that have a low carbon footprint when fed to energy storage devices.
2. EVs when connected to the grid for charging may add to the peak demand. This creates a challenge during shortage of power supply. The challenge lies in arriving at a balance between the demand and supply curves.
3. Optimized usage of charging stations is a necessity for large-scale charging keeping in mind the technology and variation in charging set-ups. For electric utilities, off peak hour charging helps augment the load curve. An example of charging stations data in Japan (Figure 10.1) shows the stark difference between normal and quick-charging stations.

No. of Charging Stations in Japan

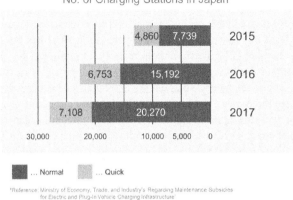

'Reference: Ministry of Economy, Trade, and Industry's Regarding Maintenance Subsidies for Electric and Plug-In Vehicle Charging Infrastructure'

FIGURE 10.1 Difference in normal and quick charging stations in Japan

4. EVs, when parked and charged can feed the power back to the grid (if suitable provisions are made) thereby supporting the utility through vehicle-to-grid power (V2G) [2]. Another means to reduce the charging at peak hours is demand response (DR), whereby the choice and incentive is given to the consumer to make a choice regarding the charging time.

5. With suitable connections, V2G allows the EVs to become distributed energy storage systems for the electrical grid. However, the electrical energy delivered back to the grid must be priced such that the additional cost incurred must be recovered back. This recovery in cost is required as the life span of the battery is reduced due to frequent charging–discharging. The distributed storage provides advantages such as making the grid more steady, secure and resilient by regulating frequency and the spinning reserve as backup power in the distribution system.

6. Due to the intermittent nature of solar and wind energy, large-scale integration can be facilitated by the V2G system. For the worldwide shift to the emerging green and sustainable energy economy, V2G is an important enabler. Compared to centralized power generation, the comparatively lesser investment costs of vehicle power systems and the small incremental expense to alter EVs to generate grid power imply economic affordability.

7. Despite the V2G arrangement, EVs will contribute a smaller amount of power because the unit cost of energy from EVs is greater than bulk electricity from centralized generators and due to their elemental engineering features. V2G comes into the fray in the case of ancillary service markets of spinning reserves and regulation, particularly when there is a capacity payment to be online and available, with an added energy payment when power is actually dispatched. In such cases, V2G power can more than make up for the money lost on each unit sold, with the capacity payment. But in peak power markets, V2G can participate when compensated only for energy, but only when power tariffs are unusually high.

10.6   SOLUTION THROUGH INTERNET OF THINGS

1. To address the challenges of green smart cities, various deep-learning approaches can be used in order to optimize the generation of power required for extra load of charging EVs. This will help reduce and minimize the operational cost and we can apply the various machine learning approaches on the prediction of power using wind energy, solar energy and other renewable energy resources.

2. To resolve the issue of load balancing in the power generation system, various machine learning approaches can be used to reduce the gap of demand and supply of load. Load estimation or forecasting can help in estimating the deficit amount or predict the amount required to meet the demand in electric power supply. This estimation requires that the charging profile of EVs constitutes the peak demand period.

3. Electricity load forecasting at various charging stations can be done with the help of numerous machine-learning algorithms with IoT integrated systems so that charging of EVs at off-peak hours supplements the load curve for electric utilities. Various setups and technologies require accurate optimization for the use of a large number of EVs.

4. V2G provision through demand response provides an incentive to the consumer to obtain information of dynamic tarrif (changing prices). Use of IOT enabled devices educates the consumer regarding the dynamic tarrif and thereby allows the V2G grid though demand response.

5. Internet of Things (IoT) is an integral part of the smart grid and the shift from traditional to smart grid and conventional to renewable sources has been a big step forward in the power industry. The advantages of the IoT integrated smart grid, in line with the sustainable goals results in reduced carbon footprint, higher energy production, a more secure grid, decreased vulnerability to any external influence, consistent power supply and a consumer centric approach. The proposed model architecture and methodology using IOT is shown in Figure. 10.2

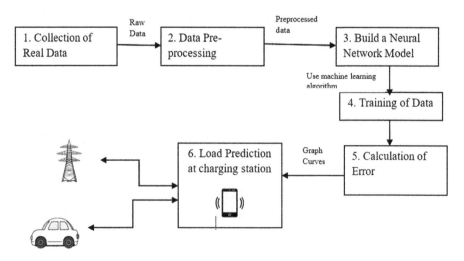

FIGURE 10.2   IOT and ML based system architecture for EV charging stations

Step 1: Collection of data: The data collection includes those related to EV charging points. The data attributes are: Number of charging stations in an area, usage/load curve, number of vehicles visited at charging station, number of charging stations in a particular area.

Step 2: Preprocessing of data: The data will be cleaned by using preprocessing techniques and then this data will be used to build a neural networks model for implementing a machine learning approach.

Step 3: Training of data: During the creation of the network model, 90% of the dataset will be used as a training dataset while 10% of it will be used for testing the model. The libraries such as Tensorflow, Keras, Pandas etc. will be used to train the models.

Step 4: Testing and calculation of error: A mean squared error is calculated for the actual value of the test set and the predicted value.

Step 5: Formulation of graph: Python libraries such as matplotlib has been used to plot the graphs between the actual value and the predicted value.

Step 6: Prediction curves: The visualization results will be made available through an android application on mobile phone of a user.

This application will reduce the waiting time of the consumer giving him a choice of minimum waiting time within an area of set charging station. An incentive-based tariff can be provided thereby distributing the load on these charging stations. This green initiative using IoT will provide a sustainable solution benefiting both the consumer and the environment.

REFERENCES

[1]   Khan, Q., Ahmad, F., and Imran, M. (2017) Congestion management in Indian power transmission system. *International Journal of Engineering Technology*, 9(3): 26–31.

[2]   Ahmad, F., and Alam, M. S. (2017) Feasibility study, design and implementation of smart polygeneration microgrid at AMU. *Sustainable Cities and Society* 35: 309–322.

[3]   Ahmad, F., Alam, M. S., and Asaad, M. (2017) Developments in xEVs charging infrastructure and energy management system for smart microgrids including xEVs. *Sustain Cities Soc,* 35: 552–564.

[4]   Wu, H., and Shahidehpour, M. (2016) A game theoretic approach to risk based optimal bidding strategies for electric vehicle aggregators in electricity markets with variable wind energy resources. *IEEE Transactions on Sustainable Energy,* 7(1): 374–385.

[5]   Asaad, M, Ahmad, F., Saad Alam, M., and Rafat, Y. (2017) IoT enabled electric vehicle's battery monitoring system. In *Proceedings of the 1st EAI International Conference on Smart Grid Assisted Internet of Things*, Sault Ste. Marie, Ontario: European Alliance for Innovation (EAI).

[6]   Sony Corporation. *Lithium Ion Rechargeable Batteries- Technical Handbook*. www.sony.com.cn/products/ ed/battery/download.pdf/. Accessed 15 February 2017.

[7]   Node-Red. JS Foundation. Flow-based programming for the Internet of Things. http://nodered.org/. Accessed 25 January 2017.

[8]   MQTT (2017) Machine-to-Machine (M2M) learning. http://mqtt.org. Accessed 20 February 2017.

[9]   HiveMQ. MQTT publish & subscribe. www.hivemq.com/blog/mqtt-essentials-part2-publish-subscribe. Accessed 25 January 2017.

[10] HiveMQ. MQTT security fundamentals. www.hivemq.com/blog/introducing-the-mqtt-securityfundamentals. Accessed 15 August 2017.

[11] WEF_2018_ Electric_For_Smarter_Cities.pdf.

[12] Li, C.-T., Ahn, C., Peng, H., and Sun, J. (2011) Integration of plug-in electric vehicle charging and wind energy scheduling on electricity grid. In *IEEE Innovative Smart Grid Technologies (ISGT)*, Washington, DC: IEEE, pp. 1–7.

[13] Aunedi, M., and Strbac, G. (2013) Efficient system integration of wind generation through smart charging of electric vehicles. In *Proceedings of 8th International Conference and Exhibition on Ecological Vehicles and Renewable Energies (EVER)*, Monte Carlo: IEEE, pp. 1–12.

[14] Jin, C., Sheng, X., and Ghosh P. (2013) Energy efficient algorithms for electric vehicle charging with intermittent renewable energy sources. In *IEEE Power and Energy Society General Meeting (PES)*, Vancouver, BC, Canada: IEEE, pp. 1–5.

[15] Bouallaga, A., Merdassi, A., Davigny, A., Robyns, B., and Courtecuisse, V. (2013) Minimization of energy transmission cost and CO2 emissions using coordination of electric vehicle and wind power (W2V). In *IEEE PowerTech (POWERTECH)*, Grenoble, France: IEEE, pp. 1–6.

[16] Battery electric vehicle (2014) http://en.wikipedia.org/ wiki/Battery_electric_vehicle.

[17] Haghbin, S., Khan, K., Lundmark, S., Alakula, M., Carlson, O., Leksell, M., et al. (2010) Integrated chargers for EV's and PHEV's: examples and new solutions. In *Proceedings of XIX International Conference on Electrical Machines (ICEM)*, Rome: IEEE, pp. 1–6.

[18] Kong, F., Dong, C., Liu, X., and Zeng, H. (2014) Blowing hard is not all we want: Quantity vs quality of wind power in the smart grid. In *Proceedings of the 33rd IEEE Conference on Computer Communications*, Toronto: IEEE, pp. 2813–2821.

[19] Kong, F., Dong, C., Liu, X., and Zeng, H. (2014) Quantity vs quality: Optimal harvesting wind power for the smart grid. *Proceedings of IEEE,* 102: 1762–1776.

[20] Muller-Steinhagen, H. and Trieb, F. *Concentrating solar power*. Stuttgart: Institute of Technical Thermodynamics, 2014 www.seia.org/sites/default/files/cspfactsheet-120223144940-phpapp01.pdf.

11 Disputes and Challenges in IoT (Internet of Things) Applications
A Literature Survey

Prachi Shori, Shalini Shori, and Aastha Gour

CONTENTS

11.1 INTRODUCTION

IoT is one of the recent advanced technologies in the field of computer science which holds embedded technology and information technology together. Today devices are more potent and enriched with new technology consisting of wearable devices that are contributing to the estimated growth of IoT.

In the case of the healthcare industry, with the help of IoT devices, we can facilitate remote health monitoring and notification of alarming situations. There is a wide range of fitness tracking devices these days starting from monitoring pulse rate and blood pressure, and wristbands equipped with all the health-monitoring features.

International Telecommunication Union (ITU) and European Research Cluster on IoT (IERC) described Internet of things (IoT) as "A dynamic global network infrastructure with self-configuring capabilities based on standard and interoperable communication protocols, where physical and virtual 'things' have identities, physical attributes, and virtual personalities, use intelligent interfaces and are seamlessly integrated into the information network" [1].

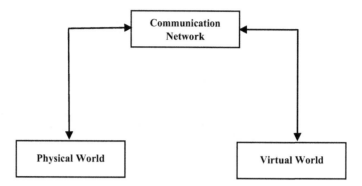

FIGURE 11.1 Basic blocks of IoT system

Dr. Raj Kamal defines IoT as:

A network of physical things (objects) sending, receiving or communicating information using the internet or other communication technologies and network just as the computers, tablets, and mobiles do, and thus enabling the monitoring, coordinating or controlling process across the internet or another data network.

[2].

Michael Miller defines IoT as "Small devices (things), each with their own Internet Protocol (IP) address, connected to other such devices (things) via the Internet" [3].

IoT is also interpreted as "An open comprehensive network of intelligent objects that have the capacity to auto-organize, share information, data, and resources, reacting and acting in face of situations and changes in the environment" [4].

The basic blocks of the IoT system consist of three components which are classified as the physical world, virtual world and communication network where the physical world such as sensors, processors, controllers and actuators interact with the virtual world (including cloud, API or mobile applications) using a communication network. A communication network is basically a network with various layers that have different functionality and protocols.

IoT is a developing concept and persists to be the cutting-edge concept in the world of computer and embedded engineering. Many types of research present the various challenges and opportunities in the field of IoT. Some of these works of literature are studied and analyzed to give an overall review of the ongoing researches in the IoT domain. This literature review is discussed in detail in the next section.

11.2 LITERATURE REVIEW

The recent researches in the field of IoT involve issues like big data analytics, data security and privacy concerns, fundamental infrastructure requirements for IoT, real-time processing and accessing of data, time complexity of IoT devices and human body interaction with the networking devices. Various researchers propose different

Count of paper published

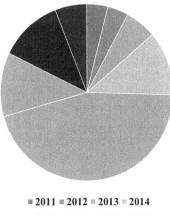

■ 2011 ■ 2012 ■ 2013 ■ 2014
■ 2015 ■ 2016 ■ 2017 ■ 2018

FIGURE 11.2   A pie-chart showing paper count versus publication year of papers used in our study from 2011–2018

solutions to the listed issues, some present various problem-solving techniques and some compare the different techniques and algorithms designed and proposed in other researches.

We cannot think of one without the other when the discussion is being made on the following: IoT and big data as they are very closely entangled.

The existence of the term big data came before IoT in the field of analytics. big data is interpreted using the four Vs: veracity, velocity, variety and volume [55].

IoT devices offer big data, from which big data analytics can mine information to generate understandings required of it.

This literature assessment covers the big data issues, security and challenges in maintaining the privacy of data, issues with healthcare data, and related challenges while handling such big data in IoT applications.

The following section presents the literature review for big data analytics, the data security and privacy issues are reviewed in section 11.2.2, the usage of IoT in healthcare systems are discussed in section 11.2.3 and in section 11.2.4 services and other related issues of IoT are discussed.

11.2.1   BIG DATA ANALYTICS

The first most challenging issue with IoT is the large amount of heterogeneous data it has to handle. A cloud storage solution [5], stores a massive amount of diverse data and delivers them in a systematic way. The cloud architecture involves a data manager, which performs the functions of data abstraction and data enrichment. REST API is provided to access the data [5] and [6]. MongoDB REST server written in Java (based on a jetty web server) [5] and communication APIs (based on the REST web server) [6] acquire and manage sensor data on the cloud. The latter architecture comprises (i) IoT devices like mobile sensors and wearables, which obtain biosignals, motion and background information; and (ii) sensor gateway, which gathers all the signals from sensors and forward them to communication APIs.

Cloud computing services can also be used for Supervisory Control And Data Acquisition (SCADA) systems [7], to provide better collaboration and reduced energy costs. This system enables the in real time and governs the power grid, and brings together the data from perception devices and sensor nodes. The data sharing in cloud storage causes increased security and network issues. This issue can be solved by using fog computing. Fog computing refers to the processing of data at the entry point of networks without any data transfer to the cloud. A fog node [8] is an entity that has the capability to compute data, store data and connect to the network. The main characteristics of fog node are: (i) connectivity, (ii) a two-way data exchange, (iii) flexible integration, (iv)protocol translation, (v) data aggregation/ filtering/ formatting and encoding/ decoding, (vi)short-time database, (vii) security and data protection, (viii) assessment and notification and (ix) local processing.

Fog-driven systems require extremely time-sensitive decisions to be made [8]. In the case of processing real-time data, big data analytics is beneficial when used with Hadoop [9], which implements significant real-time analysis on the massive volume of data and enables the system to foretell the disastrous situations before they occur in case of healthcare systems. The big data set is further subdivided into smaller chunks of data sets in Hadoop MapReduce [7], to perform data processing for static applications only, but for real-time big data processing, the most appropriate platform is stream processing as it provides high scalability, availability and fault-tolerant architecture. The stream of big data can be visualized as a massive stream [10] of small pieces of spatial-temporal data that must be gathered, stored, indexed and processed in real-time. Processing of such streamed data requires fully decentralized architectures. It raises challenges like data transportation, data processing and data cleaning in streams.

An amalgamation of statistical and declarative model-based cleaning methodology is applied for cleaning data streams. In order to purify streams of data in real time the above framework [11] is designed for the Distributed Cleaning System (DCS). The functions of this framework are (i) Point – It works on single values to identify and filter the single outliers, (ii) Smooth – Target multiple values to find missing values using temporal windowing, (iii) Merge – Identifies single-type sensors which are spatially related and reads them, (iv) Arbitrate – Identifies contradictory readings among spatial groups which are nearby and corrects them, and (v) Visualize – It combines the readings of different spatial groups or for different types of sensors.

An outlier detection algorithm [12], uses the Mahout K-means algorithm along with big data processing. This algorithm estimates values classifies and recognizes behaviors and detects outliers; and together with the LinkSmart middleware pattern services, this algorithm contributes to the decrease in network traffic in the context of IoT as the upper application layers do not receive raw data but the pre-processed information. The concept of pre-processing is adapted [13], for outlier detection which is achieved by two algorithms: distance-based outlier detection algorithm and cluster-based outlier detection algorithm. The three types of outliers such as contextual outliers, point outliers, and collective outliers can be detected and removed by applying the model, based on the presented algorithms.

A semantic data model [14] deals with the real-time data, which stores and interpret IoT data in real time. A data-accessing method (UDA-IoT) that is resource-based is intended to collect and process IoT data ubiquitously, with the semantic data model. This method comprises establishing a relationship between IoT physical entity and information entity, a mapping between the business function and transitive resources, cloud management platform and data accessing process with UDA-IoT. This real-time data must be structured and resized into nominal value to develop efficient data analytics. The predictive analysis algorithm [15], predicts the conditions associated with it even before they actually occur. The structural design of the prediction analysis system involves numerous stages, such as data collection, data warehousing, and predictive analysis processing and analyzed reports.

The categorization of big data and mining the information out of this big data is a big issue in IoT. A platform [16] designed for extracting the information allied to the IoT, includes both the sensor's data and data-about-data (meta-data). This platform comprises the following components: (i) ServIoTicy and (ii) iServe, designed for scalability. ServIoTicy stores and processes data and provides capabilities to consume the data along with the real-time transformation of data generated by sensors. "iServe" does the publication and discovery of sensors meta-data and provides the ability to use efficiently the semantic information gathered to operate sensors.

The applicability of various renowned data-mining algorithms is examined [17], on the real-time datasets. By experimental analysis using R-platform, it is found that C4.5, C5.0, ANN (Artificial Neural Network) and DLANN (Deep Learning Artificial Neural Network) can give relatively more accurate results for data mining as compared to SVM (Support Vector Machine), KNN (K-Nearest Neighbours), LDA (Linear Discriminant Analysis) and NB (Naïve Bayes) data mining algorithms.

Data mining and machine learning data-oriented association methods are limited to small-scale datasets in terms of accuracy. These associations are highly dependent on statistics instead of logic. Semantic technologies are being challenged by issues like flexibility and the dynamic environment when identifying and maintaining the associations. Technologies may be rule-oriented or ontological-oriented. The enhancement of logical associations for IoT entities are available in the literature [18]. The author presents a method to construct data objects semantically which are based on context for logical relationships using service logics and associations are created based on the similarity between them. The construction of context-centric has four factors: (i) Terminology – It is a concept that defines all the entities involved in the framework, (ii) Assertion – When terminology is noticed and detected by environment data, (iii) Role – Parameter to describe the characteristics, (iv) Rule – Lays down the body of laws that the role must follow to establish association.

Ontology OWL (Web Ontology Language) format allows information extraction from the database [19]. In this format, those records are identified that can deal with missing data and noise using meta-data modeling. A Knowledge-Based System (KBS) [20], can be constructed using the combination of OWL and Semantic Web Rule Language (SWRL). It basically integrates multiple knowledge sources and aims at problem-solving modelling. The parameters for the modeling of OWL-based KBS are: (i) analyzing problem scenario, (ii) modeling the domain ontology, (iii) modeling the task ontology and (iv) developing semantic rules.

Along with the data processing and data management issues, IoT based systems need to take care of the data security and privacy issues as well, which will be discussed in the next section.

11.2.2 SECURITY AND PRIVACY

As discussed in the previous section, data acquisition, data collection, storage of data and interpretation of useful information require various data acquisition methods, pre-processing techniques, categorization or classification techniques and various practices of data mining. But the security of the data and to maintain the privacy of confidential information is highly required when working in the IoT environment. Many of the security issues are raised in various researches.

A number of possible types of attacks against IoT devices [21], are: (1) vulnerability exploitation – buffer overflow/ code injection/ cross-site scripting, (2) password attacks, (3) malware attacks, (4) spoofing, (5) man-in-the-middle attacks/ sniffing, (6) botnet enslaving, (7) remote access and (8) data leakage. The IoT security vulnerabilities are also listed in [21], such as (1) simplicity and easy to use, (2) log in credentials that are default and not secure, (3) software updates that are below standard, (4) user interface is not secure, (5) user privacy is poor and (6) encryption of communication is lacking. These issues can be solved by adapting techniques like regularly changing the usernames and default passwords after an interval, checking the latest update of software and upgrading the software accordingly, switching to settings like login lock and two-factor authentication, proper encryption, using a secondary network for IoT devices, to keep the home Wi-Fi secure disconnecting the devices from the internet when not in use, for security tips always go through the device manual, download security applications and you can even switch to hardware solutions to protect the IoT network from outside attacks.

While integrating IoT and cloud computing, there are some features of cloud computing [22], which are to be taken care of from the security point of view. These are (i) security model of cloud computing, (ii) security issues of mobile cloud computing and (iii) trade-offs such as privacy and security, performance, latency or delay and connectivity in mobile cloud computing. Major challenges in the view of security when considering IoT and cloud computing integration are (i) performance, (ii) heterogeneity, (iii) big data, (iv) reliability and (v) monitoring.

A survey [23], discusses the targets for attacks in IoT systems, which are: (i) hardware threats, (ii) network threats and (iii) smart application threats. A few security challenges are also given in this survey, such as secure SG, lightweight authentication, heterogeneity and QoS. These security issues can be solved by using several security architectures like architecture based on Software-Defined Networking (SDN) [24]. The basic principle behind SDN architecture is to cover the endpoint devices that are accessing the network in the perimeter of security. The proposed distributed network access control architecture does two things: it keeps track of attacks taking place in the multiple SDN domains and also shares the traffic load along with helping out the root controller to manage it. To ensure network safety, the SDN controllers act as security guards on the boundary of the SDN domain. The network manager of SDN engages in QoS administration, load balancing, edge/ fog computing and traffic engineering [25].

An end-to-end security scheme [26], includes (i) DTLS handshake-based certificate for the authentication of the end-user, (ii) based on the session, end-to-end communication must be secure and (iii) smart gateways that are interconnected promote robust mobility. A smart gateway can be imagined as an intermediate processing layer called fog layer between sensors and IoT devices called a device layer, and cloud layer which consists of cloud services. The IoT based mobile-healthcare systems [27], are decentralized systems and their security concerns are: (i) authenticity, (ii) privacy, (iii) confidentiality, (iv) integrity, (v) checking authentication of data origin, (vi) authentication of entity and non-repudiation. In the proposed M-Healthcare system, Public Key Infrastructure (PKI) and PKI certificates are used as the basis of the concerns in security. There is always a possibility of extension of security issues from users to devices, then to applications and finally to services.

Distributed systems along with the interacting devices lack the support of authorization frameworks in the case of mechanisms that are effective and efficient along with being scalable and manageable. In [28], the system that can access their own services and information using their own access control processes is known as CapBAC (capability-based access control) system. The CapBAC system possesses the additional features that enable the system to provide (i) revocation of capability, (ii) support of delegation and (iii) granularity of information. The elements of CapBAC architecture are: (1) the resource object of the capability, (2) the authorization capability, (3) the capability revocation, (4) the service/ operation request, (5) the PDP (Policy Decision Point), (6) the resource manager and (7) the revocation service. The major advantages of this system are, it provides access control that is more efficient and fine-grained and has fewer security concerns.

The manual security analysis suffers from time complexity; to reduce the latency security analysis can be automated. A framework for automating security analysis [29] contains five stages: (1) processing of data, (2) building the security model, (3) visualization of security, (4) study of security and (5) updates if required in the model.

The above-discussed security methods based on encryption are old-style and are not suitable enough to protect the data in the cloud. Malicious intruders can be identified and prevented using big data analytics. A meta cloud data storage security architecture [30], stores data on the basis of scope and importance in numerous cloud data centers. This framework produces a cryptographic value of the encrypted storage path. This value is known as cryptographic virtual mapping of big data. Using the meta cloud data storage interface, this framework secures the mapped images of virtual data elements to each source. The MapReduce algorithm is used for big data analytics in the distributed cloud data center [7], [30] and it can easily work with huge sized data.

There are a variety of sets of security issues, related challenges and requirements that are presented by various researchers in their work. IoT based smart healthcare systems create a huge amount of big data that produces challenges like data security. Some other challenges and their problem-solving techniques will be discussed in the next section.

11.2.3 IoT FOR HEALTHCARE

Heterogeneous data is being produced by IoT based healthcare systems or e-health systems in huge volumes. As data is on a large scale so are the issues of data security. Body Sensor Network (BSN) has the following major requirements for security for healthcare systems [31], (i) privacy of data, (ii) freshness of data, (iii) authentication, (iv) data integrity, (v) secure localization and (vi) anonymity. These security requirements are divided into two categories: (1) Data security that consists of privacy of data, data integrity and freshness of data and (2) Network security including authentication, anonymity and secure localization. The system is able to realize data security using OCB (offset codebook) mode, which is an authentic scheme of encryption with an affordable overhead of computation.

WBASN (Web-based Body Area Sensor Network) is used to monitor and report the physiological state of the patient. It is made up of multiple sensor nodes. Integrated solution [32], based on cloud-ID comprises features like Electronic Health Records (EHR), e-prescribing system, personal health records, pharmacy systems, clinical decision systems, etc. Services like [33] PaaS (Platform as a Service) and IaaS (Infrastructure as a Service) are offered by the cloud service provider to host healthcare applications based on cloud-IoT.

A platform "EcoHealth," a web middleware platform [33], integrates the data obtained from the heterogeneous sensor and provides a mechanism to operate in real-time like monitoring, processing, visualizing and storing, then finally sending alerts about the patients' conditions and vital symptoms. The "EcoHealth" platform is organized in several loosely coupled modules such as: (i) devices connection module, (ii) visualization and management module, (iii) storage module and (iv) common services. These modules use the JAAS (Java Authentication and Authorization Services) services offered by the JBOSS server and facilitates the easier management of large data streams and distributed components, governing user confidentiality, integrity and authenticity.

An intelligent home-based platform [34], involves (i) iMedBox, which is an intelligent medicine box and is based on open-platform, (ii) iMedPack, an intelligent pharmaceutical packaging enriched with RFID enabled communication capability, functional materials enabling the actuation capability and a bio-patch (a sensor device that is bio-medical and wearable consisting of a system chip and advanced printing technology). The network architecture of this platform involves three network layers: (1) sensor data collecting layer, (2) medical resource management layer, and (3) smart medical service layer. Using the Raspberry Pi II platform, another healthcare system is implemented [35]. Three components of the system architecture are (1) Biosensors, (2) LPU and (3) Body Sensor Network (BSN) server. The communication channels of this system are: "Sensors to LPU" and "LPU to BSN".

For the purpose of modeling and solving real-time problems, there is a powerful approach and the platform is known as Multi-Agent-System (MAS) [36]. MAS is a loosely coupled network of problem-solver entities or agents, which work in a uniform manner to discover solutions to problems that are beyond the reach or knowledge of individual entities. Wireless Sensor Network (WSN) and Body Area Sensor

Network (BASN) are the specific technologies well suited for modeling MAS. The proposed MAS architecture involves: (i) external enabler level – acts as a bridge to interface with many different networks, (ii) core MAS platform level – supports the creation, launching, reception and execution of stationary and mobile agents and their interactions and (iii) vertical services level – provides functionality for other services like web service access.

A Multi-Agent Distributed Information Platform (MADIP) [37] and [39], comprises six components starting from (1) resource agent, (2) physician network, (3) user agent, (4) knowledge-based data server, (5) diagnostic agent and (6) external services. The heterogeneous e-Health environment [37] and [38], suffers from issues like scalability, openness and interoperability, which are henceforth handled by Java Agent Development (JADE) framework. The hosts of the framework [38] work in a runtime environment. An AMS agent (Agent Management System) is authorized for creation and termination of remote agents and a DF (Directory Facilitator) agent enables an agent to find other agents to provide the required services.

The IoT based telehealth system [40] captures the crucial signs and delivers the required data care remotely using a wearable ECG wireless sensor. This type of technology: (i) helps in the identification of the earlier symptoms of health problems, (ii) in acute situations the healthcare providers are informed, (iii) discover a balance between health and lifestyle and (iv)isolated location must be provided with the healthcare facilities. Ontology [41], based Automating Design Methodology (ADM) comprises (1) real-time detection of a patient's biosignals using smart monitoring devices, (2) establishment of prototype rehabilitation system facilitating all the main treatments, (3) constant monitoring of patients, nurses or doctors through Internet Protocol (IP) camera and (4) analysis of data along with forwarding the sensed data is carried out by a home gateway. In the framework of IoT, automating design methodology involves the four phases of the process: Phase I involves stages like diagnosis along with evaluation, Phase II – working on scheme design of rehabilitation, Phase III – working on the design of subsystem and finally Phase IV – detail design.

Rather than using medical sensors, distributed smart e-health gateways can be used to perform authentication and authorization of a remote end-user. An architecture [42], presents DTLS (Datagram Transport Layer Security), which is a handshake protocol and is certificate-based. It is regarded as the primary IP security solution for IoT. This architecture comprises smart e-health gateway, Medical Sensor Network (MSN), web-clients and back-end system. The healthcare system architecture [43], comprises medical sensors and actuators network, networks of smart e-health gateways and back-end system. Inter device communication is taken care of by smart e-health gateways at the fog layer along with facilitating the various wireless protocols support.

There are three units in IoT based quality-aware ECG monitoring system [44], (i) the signal sensing module known as ECG, (ii) another module is automated signal quality assessment (SQA), and (iii) signal quality-aware (SQAw) ECG analysis and transmission module. A lightweight ECG SQA method is presented in this research automatically classifies the received ECG signals into two classes: acceptable and un-acceptable class. The proposed framework is capable of performing: (1) baseline wander removal and abrupt change detection, (2) ECG signal absence detection, (3) High Frequency (HF) noise detection and (4) ECG signal quality grading.

- Wireless Sensor Network (WSN) and Radio Frequency IDentification (RFID) along with smart mobile interoperability with REST network infrastructure [45], or Multi-Agent IoT [46], are incorporated in the smart healthcare systems. The RFID technology is a low-cost solution and is energy-autonomous to disposable sensors [47]. A passive RFID system comprises a digital device namely tag, embedding an IC-chip having a unique ID-code and an antenna together with a reader which is a radio scanner device. The technologies RFID and IoT integrated together within a system enables a ubiquitous and quick access to healthcare data. The prototype [48], based on Electronic Product Code (EPC) and Object Name Service (ONS), enables ubiquitous and quick access to healthcare data.

11.2.4 SERVICES AND RELATED ISSUES OF IoT

The IoT services can be categorized [49] according to the technical features such as (1) identify-related services (Active/ Passive), (2) information aggregation services, (3) collaborative-aware services, which require "terminal-to-terminal" as well as "terminal-to-person" communication and (4) ubiquitous services. As IoT deals with the real-time data, the architectures and applications designed for real-time processing are of great interest. To be able to detect dynamic changes in real time, a context-sensitive system [50] is realized using a cyber-physical system (CPS) that integrates sensors with cyber components. The primary CPS element has the job of collecting the raw data from the dynamic physical environments, then analyzing the data and converting it into useful knowledge. In order to extract the necessary information cluster analysis algorithm of machine learning can be used.

Self-Organizing Map (SOM) methodology [51], is an efficient tool used for visualizing the data as it performs clustering and vector quantization on the dataset. SOM is a time complex tool so not as attractive for quickly mining information from big data. The scheme by offering clustering and data proximity technique is able to emulate SOM. It has three phases: (i) data summarization into prototypes, done by adapting the first level of k-means for vector quantization; (ii) clustering of prototypes, is done by the second level of k-means for prototype clustering; (iii) data mapping is done by anchor projection mapping via multidimensional scaling.

A fully distributed Efficient Application-layer Discovery Protocol (EADP) [52], discovers services in IP-based ubiquitous sensor networks. This protocol is able to provide ideal acquisition times, less consumption of energy and lesser overhead. It uses an adaptive pull-push model. The EADP architecture consists of three main elements, which are UA, SA and state-maintenance mechanism. The task of UA (User-Agent) is during the pull phase to discover services. SA (Service Agent) basically registers, advertises and does service maintenance during the push phase. The last phase, which is the state-maintenance mechanism, is in charge of making the protocol react seamlessly to any topology changes.

One has to deal with the numerous challenges [53], while performing access control in IoT, such as (i) a large number of connected systems and devices, (ii) access control policies are made too complex due to the dynamism of IoT devices, (iii) a fine-grained access of access control must be available to the groups, (iv) there should be flexibility in mechanism of access control, (v) uncomplicated devices and restrained-resources should also be supported by the mechanism and (vi) to provide a user-friendly UI to both the consumers and devices. Security and privacy are the key objectives of access control framework [54], which are defined in terms of: (i) transparency, (ii) user-driven access, (iii) anonymity, (iv) pseudonymity, (v) unlikability, (vi) unobservability and (vii) decentralization. The technologies constraints are flexibility, scalability, lightweight and heterogeneity. The social and economic aspects of IoT for access control are availability, usability, confidentiality and integrity and reliability.

11.3 CONCLUSION

In real-time IoT applications, the interfacing of software and hardware is a big challenge. The data generated by physical sources are very big in size and heterogeneous in nature. To handle and process such data, big data analytics can be a good solution as suggested by the reviewed researches. Different architectures and technologies (like RFID, MAS) are proposed and developed to capture the data from sensors and to pre-process this raw data. Some technologies are adapted to store such big data (like cloud computing). Various classification methods (like clustering) are designed to categorize the data into suitable groups then to extract useful information and various algorithms are applied (like a k-means algorithm). Various frameworks and platforms are developed to resolve issues like security, data privacy, time complexity, accuracy, adaptability to work in real-time.

There are some issues like secure data storage and processing in real time with a significant accuracy, latency between data processing and inferring the desired information, cost-effective solution for biosensors and other infrastructure to develop a pocket-friendly IoT systems, need to be worked upon, and these fields are among the ongoing researches in IoT all over the world.

REFERENCES

[1] Vermesan, O., and Friess, P. (2014) *Internet of Things – from Research and Innovation Market Deployment,* ISBN: 978-87-93102-94-1, Aalborg: River Publishers.
[2] Kamal, R. (2017) *Internet of Things – Architecture and Design Principles.,* New York: McGraw-Hill Education.
[3] Miller, M. (2015) *The Internet of Things: How Smart TVs, Smart Cars, Smart Homes and Smart Cities are Changing the World,* ISBN-13: 978-0-7897-5400-4, Indianapolis: Que publishing.
[4] Madakam, S., Ramaswamy, R., and Tripathi, S. (2015) Internet of Things (IoT): A literature review. *Journal of Computer and Communications*, 3(05), 164.
[5] Fazio, M., Celesti, A., Puliafito, A., and Villari, M. (2015) Big data storage in the cloud for smart environment monitoring. *Procedia Computer Science*, 52, 500–506.

[6]   Doukas, C., and Maglogiannis, I. (2012) Bringing IoT and cloud computing towards
      pervasive healthcare. In *Innovative Mobile and Internet Services in Ubiquitous
      Computing (IMIS), 2012 Sixth International Conference,* July, Palermo, Italy: IEEE,
      pp. 922–926.

[7]   Jaradat, M., Jarrah, M., Bousselham, A., Jararweh, Y., and Al-Ayyoub, M. (2015) The
      internet of energy: Smart sensor networks and big data management for smart grid.
      Procedia Computer Science, 56, 592–597.

[8]   Farahani, B., Firouzi, F., Chang, V., Badaroglu, M., Constant, N., and Mankodiya, K.
      (2018) Towards fog-driven IoT eHealth: Promises and challenges of IoT in medicine
      and healthcare. *Future Generation Computer Systems,* 78, 659–676.

[9]   Archenaa, J., and Anita, E. M. (2015) A survey of big data analytics in healthcare and
      government. *Procedia Computer Science,* 50, 408–413.

[10] Cortés, R., Bonnaire, X., Marin, O., and Sens, P. (2015). Stream processing of healthcare
      sensor data: Studying user traces to identify challenges from a big data perspective.
      Procedia Computer Science, 52, 1004–1009.

[11] Gill, S., and Lee, B. (2015). A framework for distributed cleaning of data streams.
      Procedia Computer Science, 52, 1186–1191.

[12] Souza, A. M., and Amazonas, J. R. (2015) An outlier detect algorithm using big data pro-
      cessing and internet of things architecture. *Procedia Computer Science,* 52, 1010–1015.

[13] Christy, A., Gandhi, G. M., andVaithyasubramanian, S. (2015) Cluster based outlier
      detection algorithm for healthcare data. *Procedia Computer Science,* 50, 209–215.

[14] Xu, B., Da Xu, L., Cai, H., Xie, C., Hu, J., and Bu, F. (2014) Ubiquitous data accessing
      method in IoT-based information system for emergency medical services. *IEEE
      Transactions on Industrial Informatics,* 10(2), 1578–1586.

[15] Eswari, T., Sampath, P., and Lavanya, S. (2015) Predictive methodology for diabetic
      data analysis in big data. *Procedia Computer Science,* 50, 203–208.

[16] Villalba, Á., Pérez, J. L., Carrera, D., Pedrinaci, C., and Panziera, L. (2015) servIoTicy
      and iServe: A scalable platform for mining the IoT. *Procedia Computer Science,* 52,
      1022–1027.

[17] Alam, F., Mehmood, R., Katib, I., and Albeshri, A. (2016) Analysis of eight data mining
      algorithms for smarter Internet of Things (IoT). *Procedia Computer Science,* 98,
      437–442.

[18] Xiao, B., Kanter, T., and Rahmani, R. (2015) Constructing context-centric data objects
      to enhance logical associations for IoT entities. *Procedia Computer Science,* 52,
      1095–1100.

[19] Kumar, V. (2015) Ontology based public healthcare system in Internet of Things (IoT).
      Procedia Computer Science, 50, 99–102.

[20] Chi, Y. L., Chen, T. Y., and Tsai, W. T. (2015) A chronic disease dietary consultation
      system using OWL-based ontologies and semantic rules. *Journal of biomedical inform-
      atics,* 53, 208–219.

[21] Mohammed, A. F. (2017) Security issues in IoT. *IJSRSET,* 3(8).

[22] Stergiou, C., Psannis, K. E., Kim, B. G., and Gupta, B. (2018) Secure integration of IoT
      and cloud computing. *Future Generation Computer Systems,* 78, 964–975.

[23] Alaba, F. A., Othman, M., Hashem, I. A. T., and Alotaibi, F. (2017) Internet of Things
      security: A survey. *Journal of Network and Computer Applications,* 88, 10–28.

[24] Olivier, F., Carlos, G., and Florent, N. (2015) New security architecture for IoT network.
      Procedia Computer Science, 52, 1028–1033.

[25] Chin, W. S., Kim, H. S., Heo, Y. J., and Jang, J. W. (2015) A context-based future net-
      work infrastructure for IoT services. *Procedia Computer Science,* 56, 266–270.

[26] Moosavi, S. R., Gia, T. N., Nigussie, E., Rahmani, A. M., Virtanen, S., Tenhunen, H., and Isoaho, J. (2016) End-to-end security scheme for mobility enabled healthcare Internet of Things. *Future Generation Computer Systems*, 64, 108–124.

[27] Santos, A., Macedo, J., Costa, A., and Nicolau, M. J. (2014) Internet of things and smart objects for M-health monitoring and control. *Procedia Technology*, 16, 1351–1360.

[28] Gusmeroli, S., Piccione, S., and Rotondi, D. (2013) A capability-based security approach to manage access control in the internet of things. *Mathematical and Computer Modelling*, 58(5–6), 1189–1205.

[29] Ge, M., Hong, J. B., Guttmann, W., and Kim, D. S. (2017) A framework for automating security analysis of the internet of things. *Journal of Network and Computer Applications*, 83, 12–27.

[30] Manogaran, G., Thota, C., and Kumar, M. V. (2016) MetaCloudDataStorage architecture for big data security in cloud computing. *Procedia Computer Science*, 87, 128–133.

[31] Gope, P., and Hwang, T. (2016) BSN-Care: A secure IoT-based modern healthcare system using body sensor network. *IEEE Sensors Journal*, 16(5), 1368–1376.

[32] Tyagi, S., Agarwal, A., and Maheshwari, P. (2016) A conceptual framework for IoT-based healthcare system using cloud computing. In *Cloud System and Big Data Engineering (Confluence), 2016 6th International Conference*, January, Noida India: IEEE, pp. 503–507.

[33] Maia, P., Batista, T., Cavalcante, E., Baffa, A., Delicato, F. C., Pires, P. F., and Zomaya, A. (2014). A web platform for interconnecting body sensors and improving health care. *Procedia Computer Science*, 40, 135–142.

[34] Yang, G., Xie, L., Mäntysalo, M., Zhou, X., Pang, Z., Da Xu, L., … and Zheng, L. R. (2014) A health-IoT platform based on the integration of intelligent packaging, unobtrusive biosensor, and intelligent medicine box. *IEEE transactions on industrial informatics*, 10(4), 2180–2191.

[35] Yeh, K. H. (2016) A secure IoT-based healthcare system with body sensor networks. *IEEE Access*, 4, 10288–10299.

[36] Shakshuki, E., and Reid, M. (2015) Multi-agent system applications in healthcare: Current technology and future roadmap. *Procedia Computer Science*, 52, 252–261.

[37] Reid, M., and Shakshuki, E. M. (2017) Implementing a multi-agent system for recording and transmitting biometric information of elderly citizens. *Procedia Computer Science*, 113, 334–343.

[38] Su, C. J., and Wu, C. Y. (2011) JADE implemented mobile multi-agent based, distributed information platform for pervasive health care monitoring. *Applied Soft Computing*, 11(1), 315–325.

[39] Benhajji, N., Roy, D., and Anciaux, D. (2015) Patient-centered multi agent system for health care. *IFAC – PapersOnLine*, 48(3), 710–714.

[40] Raad, M. W., Sheltami, T., and Shakshuki, E. (2015) Ubiquitous tele-health system for elderly patients with Alzheimer's. *Procedia Computer Science*, 52, 685–689.

[41] Fan, Y. J., Yin, Y. H., Da Xu, L., Zeng, Y., and Wu, F. (2014) IoT-based smart rehabilitation system. *IEEE transactions on industrial informatics*, 10(2), 1568–1577.

[42] Moosavi, S. R., Gia, T. N., Rahmani, A. M., Nigussie, E., Virtanen, S., Isoaho, J., and Tenhunen, H. (2015) SEA: A secure and efficient authentication and authorization architecture for IoT-based healthcare using smart gateways. *Procedia Computer Science*, 52, 452–459.

[43] Rahmani, A. M., Gia, T. N., Negash, B., Anzanpour, A., Azimi, I., Jiang, M., and Liljeberg, P. (2018) Exploiting smart e-health gateways at the edge of healthcare

internet-of-things: A fog computing approach. *Future Generation Computer Systems*, 78, 641–658.

[44]   Satija, U., Ramkumar, B., and Manikandan, M. S. (2017) Real-time signal quality-aware ECG telemetry system for IoT-based health care monitoring. *IEEE Internet of Things Journal*, 4(3), 815–823.

[45]   Catarinucci, L., De Donno, D., Mainetti, L., Palano, L., Patrono, L., Stefanizzi, M. L., and Tarricone, L. (2015) An IoT-aware architecture for smart healthcare systems. *IEEE Internet of Things Journal*, 2(6), 515–526.

[46]   Turcu, C. E., and Turcu, C. O. (2013) Internet of things as key enabler for sustainable healthcare delivery. *Procedia – Social and Behavioral Sciences*, 73, 251–256.

[47]   Amendola, S., Lodato, R., Manzari, S., Occhiuzzi, C., and Marrocco, G. (2014) RFID technology for IoT-based personal healthcare in smart spaces. *IEEE Internet of Things Journal*, 1(2), 144–152.

[48]   Laranjo, I., Macedo, J., and Santos, A. (2012) Internet of things for medication control: Service implementation and testing. *Procedia Technology*, 5, 777–786.

[49]   Gigli, M., and Koo, S. G. (2011) Internet of Things: Services and applications categorization. *Advance in Internet of Things*, 1(2), 27–31.

[50]   Spezzano, G., and Vinci, A. (2015) Pattern detection in cyber-physical systems. *Procedia Computer Science*, 52, 1016–1021.

[51]   Cordel II, M. O., and Azcarraga, A. P. (2015) Fast emulation of self-organizing maps for large datasets. *Procedia Computer Science*, 52, 381–388.

[52]   Djamaa, B., & Witty, R. (2013). An efficient service discovery protocol for 6LoWPANs. In *Science and Information Conference (SAI), October 2013* London: IEEE, pp. 645–652.

[53]   Adda, M., Abdelaziz, J., Mcheick, H., and Saad, R. (2015). Toward an access control model for IOTCollab. *Procedia Computer Science*, 52, 428–435.

[54]   Ouaddah, A., Mousannif, H., Elkalam, A. A., and Ouahman, A. A. (2017) Access control in the Internet of things: Big challenges and new opportunities. *Computer Networks*, 112, 237–262.

[55]   Ravindra, S. (2017) Understanding the relationship between IoT and big data. *JAX Magazine*, 7(4), 15–22.

12 Security and Privacy for a Green Internet of Things

Anand Mohan

CONTENTS

The Internet of Things (IoT) will control the critical infrastructure of the twenty-first century, including smart cities, smart manufacturing, the smart power grid, and smart transportation systems. These smart cyberphysical systems will interconnect billions of smart devices with IoT control systems, thus requiring an IoT with exceptionally low latencies and exceptionally high cybersecurity

12.1   INTRODUCTION

The US National Academy of Engineering recently identified 14 grand challenges for the twenty-first century, including achieving cybersecurity for the IoT. Fundamentally new approaches to achieving cybersecurity, privacy, and trust in the IoT are needed that go well beyond current approaches. Figure 12.1a illustrates the US electrical power grid, which includes 7,000 power plants and about 3 million miles of transmission lines.

Figure 12.1b shows the US oil and gas pipeline network, which includes about half a million miles of oil and gas pipelines. The cyber-physical control systems for these critical resources will include millions of smart sensors and actuators reporting to an IoT control center potentially thousands of miles away. A cyberattack against these infrastructures could have catastrophic consequences. IoT control systems must therefore support ultra-low latencies, ultra-low packet loss rates, and exceptionally strong cybersecurity and privacy. Unfortunately, today's best-effort IoT

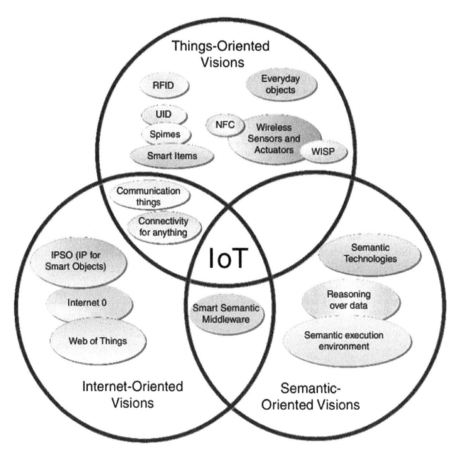

FIGURE 12.1   IoT applications

(BE-IoT) suffers from severe weaknesses. For example, it has no inherent admission control or rate control for billions of IoT users. Any user can send data at any rate to any destination at any time. Consequently, the BE-IoT suffers from frequent congestion, excessive delays, excessive packet loss, and poor energy efficiency. Delays approaching hundreds of milliseconds occur often, and denial-of-service (DoS) attacks are easy to create. The BE-IoT also offers no guarantees that data will be delivered within a strict delay deadline, or at all. According to General Electric, the future IoT could control roughly US$82 trillion in global GDP by 2030 [1]. According to Juniper Research, DoS and cyberattacks could cost global industries more than $2 trillion by the year 2020. Several international efforts are exploring ways to improve the BE-IoT to support machineto-machine and device-to-device communications. The Industrial Internet Consortium (IIC) consists of more than 250 companies and is developing a new Industrial Internet to interconnect smart factories and industrial machines [2]. The IEEE and ITU are developing a new Tactile Internet with very low latencies for human-to-machine communications [3–4]. Neither effort has considered deterministic communications, which can be inherently NP-hard. A unified, ultra-low-latency IoT network could be called the Industrial and Tactile IoT, and new approaches to achieving improved security and privacy in this Industrial-Tactile IoT are required.

Here, a new approach to achieving exceptional security and privacy in the IoT is explored. The combination of a centralized control plane using software-defined networking (SDN) technologies, the use of deterministic virtual networks (DVNs), and lightweight encryption with long keys in layer 2 can offer significant benefits.

Security and privacy in the BE-IoT A 2015 ACM SIGCOMM workshop has outlined six basic questions on the performance and security of today's BE-IoT that are difficult or impossible to answer:

- Can packets from source A reach destination B?
- Is traffic from source A and from source B isolated?
- What causes mysterious packet losses?
- Is poor performance of a cloud service caused by the BE-IoT network or the datacenter?
- Why is backbone utilization so poor?
- Is the load balancer distributing the load evenly? It is extremely difficult to provide cybersecurity, privacy, and trust in today's BE-IoT when such basic questions cannot be answered. These performance problems also incur excess capital and energy costs of tens of billions per year.

To compound the security problems, the lack of centralized control allows any user to transmit data to any destination at any data rate; thus, it is relatively easy to create a DoS attack that floods a destination with unwanted and useless traffic, causing it to fail. In 2016, a French webhosting provider, OVH, was targeted with a record-breaking DoS attack exceeding 1 Tbit per second, launched from more than 150,000 compromised smart devices, such as IP-based video cameras. Furthermore, cyberattackers are free to roam today's BE-IoT looking for targets to attack. The proposed approach will remove these threats.

12.2 BACKGROUND

The main idea of IoT is to make everyday objects such as vehicles, refrigerators, medical equipment, and general consumer goods equipped with tracking and sensing capabilities. When this vision becomes completed, "things" will also have more sophisticated networking and processing capabilities that will enable them to perceive their surroundings and interact with people. Establishing the IoT concept in the real world is possible through the integration of various technologies. Figure 12.1 summarizes the enabling technologies of IoT. Some of these technologies and trends are introduced in the next subsections, whereas some are introduced throughout the chapter.

12.2.1 Identification, Sensing and Communication Technologies

Identification is critical for IoT success. It enables IoT to name and match services with its request; therefore, billions of devices can be uniquely identified and remotely controlled over the internet. All the objects, already connected or going to be connected, must be identified by their unique identification, functionalities, and location. On IoT identification, several methods are available such as ubiquitous codes (uCode) and EPC. Addressing methods include IPv4 and IPv6. In the current IPv4, a group of coexistence sensors can be identified geographically but not individually. IPv6 can identify things globally. Furthermore, the internet mobility attributes in IPv6 may mitigate some of the device identification problems; nevertheless, the heterogeneous nature of wireless objects, concurrent operations, variable data types, and convergence of data from devices aggravate the problem.

To make IPv6 addressing suitable for lowpower wireless networks, LoWPAN is provided, which is a compression mechanism over IPv6 headers. Sensing on IoT means collecting data from related objects within the network and sending them back to the database, data warehouse, or cloud. The collected data are then analyzed to take specific actions based on the required services. The communication technologies in IoT connect heterogeneous things/objects to each other to deliver specific smart services. IoT objects should operate on low power even in the presence of lossy and noisy communication links. WiFi, Bluetooth, IEEE 802.15.4, Z-wave and long-term evolution (LTE) advanced are examples of communication protocols used in IoT. RFID, NFC and ultra-wide bandwidth (UWB) are some specific communication technologies that are also in use in IoT communication.

IoT cybersecurity is an unavoidable issue that must be solved while developing the IoT. If cybersecurity is not well managed, the hackers will take advantage of the weaknesses and defects of IoT objects and then will distort data or disrupt systems through the global IoT network. IoT failures and attacks may outweigh any of its benefits and it would be unlikely that the involved stakeholders will adopt IoT solutions widely. Hence, new techniques and methodologies have to be developed to meet IoT security, privacy and reliability requirements.

There are various techniques and devices used in IoT, such as smartphones, bar codes, cloud computing, and social networks, which to some extent affect IoT cybersecurity. At the earlier stage of IoT deployments, for example, based on RFID only, security solutions were mostly placed in an ad hoc way. From an open eco-system perspective, where different stakeholders can participate in a particular application scenario, a number of security challenges do arise, for example, one of the stakeholders owns the physical sensors/actuators, the other stakeholder handles and processes the data, and various stakeholders provide different services to the end users based on such data. Cybersecurity ensures that IoT will be a safe network for people, things, processes, hardware, and software. Cybersecurity aims to provide the world with a higher level of confidentiality, integrity, availability, scalability, accessibility, and interoperability in IoT global network.

In the upcoming years, cybersecurity issues will be one of the main research challenges of IoT. There are many reasons that make the IoT more vulnerable to attacks. The most important reasons are as follows: IoT's components spend most of their time unchecked, which makes them more vulnerable to physical attacks; since most of the communications in IoT are wireless, eavesdropping would be very simple; traditional security models cannot be adapted to new security challenges and the era of massive information produced by IoT; traditional security mechanisms and protocols are not appropriate for IoT devices because they were designed for the existing devices, which have low levels of interoperability, scalability and integrity; and IoT components mostly have low power and computing capabilities; hence, they cannot support complex security schemes.

For the IoT to be successful and to provide interoperability, reliability, effectiveness, and compatibility of the operations in a secured global scale, there is a need for security standardization at various levels. Each connected device in the IoT could be used as a doorway into the personal data or the IoT infrastructure. Moreover, the IoT introduces a new level of potential risks since mashups, interoperability, and autonomous decision making launch potential vulnerability and security loopholes. The risk of privacy will arise in the IoT because the complexity may create more service-related security vulnerabilities.

Standardization efforts in support of the IoT LHT 42 38,1 date of birth, location, budgets, etc., which is considered as one of the big data challenges. Security at both service applications and physical devices is important to the operation of IoT. Open issues remain in some areas, for example, network protocols, identity management, standardization, trusted architecture, and security and privacy protection. Security must be addressed throughout the IoT lifecycle from the initial design to running services. Figure 12.1 shows the main security challenges in IoT, which include data confidentiality, privacy, and trust. Data confidentiality is a key issue in IoT scenarios. Data confidentiality shows that data can be accessed and modified only by the authorized entities. The data of IoT applications will be linked to the physical realm; thus, ensuring data confidentiality will be essential to many use cases. Moreover, in IoT applications, the data may not only be accessed by users but also by authorized objects. Therefore, defining the access control mechanism and the process of object authentication must be addressed.

Privacy determines the rules by which data that refer to the individuals can be accessed. Privacy is a real issue that may limit the evolution of IoT. The lack of appropriate mechanisms to ensure the privacy of personal and/or sensitive information will reduce the adoption of IoT technologies. The main reason, which makes privacy a prerequisite for IoT, lies in the fact that IoT is expected to be used in critical applications like healthcare. Moreover, the risk of violation is increased because of using wireless channels, which make the system vulnerable to attack and eavesdropping due to remote access capabilities. Whereas privacy issues in traditional internet arise mostly from internet users (individuals who play an active role), privacy issues in IoT arise even from people who do not use IoT services. Accordingly, individuals must be able to determine which of their personal data can be collected, who collects such data and when these data can be collected. Moreover, the collected data should be used only when they are urgently needed to support services authorized by the accredited service providers. Trust is a complex idea used in a wide variety of contexts. Depending on the adopted perspective, different definitions of trust are possible. The main problem in many approaches to defining trust is that these definitions do not create metrics and evaluation standards of trust. Trust is considered a very important concept of IoT, even if the dynamic and fully distributed nature of IoT makes it very difficult to deal with the trust challenges.

In the IoT environment, where smart objects take the decisions themselves, the first trust relationship must be set up between humans and the objects around them. Researchers classified IoT attacks into eight categories, as shown in Figure 12.2. They also thoroughly surveyed the important aspects of IoT cybersecurity, especially the discussion of the possible four-layered IoT cybersecurity infrastructure, the major countermeasures against IoT attacks, the threats regarding application industries and the state-of-the-art current situation and possible future directions. As an open ecosystem that integrates diverse research areas, IoT is very vulnerable to attacks against security, privacy, availability, and service integrity. In [10–12], the authors modeled the IoT as a four-layer architecture: the sensing layer, the network layer, the service

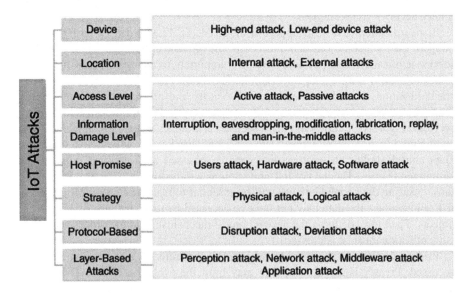

FIGURE 12.2   IoT attacks

layer and the application interface layer. At the lower layer, the sensing layer, the devices cannot provide well security protection because they have very limited energy supply and computation capacity. At the middle layers (the service layer and network layer), the network is prone to an interception, eavesdropping and DoS attacks. Furthermore, a self-organized topology, without centralized control in the network layer, can cause attacks against authentication, for example, node suppression, node replication and node impersonation.

At the upper layer, the application layer, the vulnerability problems of all layers can be mitigated using data aggregation and encryption. To build a trustworthy IoT system, a self-adaptive security policy framework and system-level security analytics are needed. Each layer is able to provide corresponding security controls such as device authentication, confidentiality in transmission, access control, data integrity and availability and the ability to prevent attacks. Furthermore, the security requirements on each layer might be different due to its features. The IoT concept was coined depending on the RFID identification and tracking techniques. Authentication is necessary to prevent data attacks between two things on the internet. As the RFID technique joined to create a large sensor network, policy questions arise, which make the security not just a technical issue. RFID cybersecurity measures include data encryption, access control, cryptography technology schemes, IPSec-based security channels and physical cybersecurity schemes.

Another enabling technology is the WSN, in which attackers can actively and aggressively attack WSN-related data or things. Hence, many protection measures have to be taken to deal with different attacks in WSN. These measures include the following: key management, in which the security keys will be generated and updated and the proper algorithm can be built; secret key algorithms, which include symmetric key and asymmetric key algorithms; security routing protocol that includes the SPINS security framework protocol, which is widely used in secure routing technologies and encompasses the Secure Network Encryption Protocol, and Micro Timed Efficient Streaming Loss-tolerant Authentication Protocol; authentication and access control; and physical security design, which covers the design of the nodes and the antennas.

The integration of WSNs and RFID enables the implementation of IoT in industrial services and the further deployment of services in more extended applications. Nevertheless, the security issues in this integration encompass the following challenges: privacy of RFID devices and WSNs devices; identification and authentication, which have to be well protected from tracking by an unauthorized user; communication security, the communication between IoT devices and RFID devices introduces security threats, which need to be addressed proactively by implementing the appropriate measures well; trust and ownership, the authenticity and integrity of the communicating things such as sensor nodes and RFID tags; integration; and user authentication.

12.3   SECURITY REQUIREMENTS

In IoT, each connected device could be a potential doorway into the IoT infrastructure or personal data [14]. The data security and privacy concerns are very important but the potential risks associated with the IoT will reach new levels as interoperability, mashups, and autonomous decision making begin to embed complexity, security loopholes, and potential vulnerability. Privacy risks will arise in the IoT since the complexity may create more vulnerability that related to the service. In IoT, much information is related with our personal information, such as date of birth, location, budgets, etc. This is one aspect of the big data challenging, and security professions will need to ensure that they think through the potential privacy risks associated with the entire data set. The IoT should be implemented in a lawful, ethical, socially, and politically acceptable way, where legal challenges, systematic approaches, technical challenges, and business challenges should be considered. This chapter focuses on the technically implementation design of the security IoT architecture. Security must be addressed throughout the IoT lifecycle from the initial design to the services running. The main research challenges in IoT scenario include the data confidentiality, privacy, and trust, as shown in Figure 12.1. To illustrate the security requirements in IoT well, we modelled the IoT as four-layer architecture: sensing layer, network layer, service layer, and application-interface layer. Each layer is able to provide corresponding security controls, such as access control, device authentication, data integrity and confidentiality in transmission, availability, and the ability to anti-virus or attacks. The most security concerns in IoT are summarized. The security requirements depend on each particularly sensing technology, networks, layers, and have been identified in the corresponding sections.

12.4 SECURITY REQUIREMENTS IN IOT ARCHITECTURE

A critical requirement of IoT is that the devices must be interconnected, which enables it to perform specific tasks, such as sensing, communicating, information processing, etc. The IoT is able to acquire, transmit, and process the information from the IoT end-nodes (such as RFID devices, sensors, gateway, intelligent devices, etc.) via network to accomplish highly complex tasks. The IoT should be able to provide applications with strong security protection (e.g. for online payment application, the IoT should be able to protect the integrity of payment information). The system architecture must provide operational guarantees for the IoT, which bridges the gap between the physical devices and the virtual worlds. In designing the framework of IoT, the following factors should be taken into consideration: technical factors, such as sensing techniques, communication methods, network technologies, etc.; security protection, such as information confidentiality, transmission security, privacy protection, etc.; business issues, such as business models, business processes, etc.

Currently, the SoA has been successfully applied to IoT design, where the applications are moving towards service-oriented integration technologies. In the business domain, the complex applications among diverse services have been appearing. Services reside in different layers of the IoT such as: sensing layer, network layer, services layer, and application-interface layer. The services-based application will heavily depend on the architecture of IoT. Figure 12.2 depicts a generic SoA for IoT, which consists of four layers:

1. sensing layer is integrated with end components of IoT to sense and acquire the information of devices;
2. network layer is the infrastructure to support wireless or wired connections among things;

3. service layer provides and manages services required by users or applications; and

4. application-interfaces layer consists of interaction methods with users or applications. The security requirements on each layer might be different due to its features.

In general, the security solution for the IoT considers the following requirements: sensing layer and IoT end-node security requirements; network layer security requirements; service layer security requirements; application-interface layer security requirements; the security requirements between layers; and security requirements for services running and maintenance.

12.4.1 SENSING LAYER AND IoT END-NODES

The IoT is a multi-layer network that inter-connects devices for information acquisition, exchange, and processing. At the sensing layer, the intelligent tags and sensor networks are able to automatically sense the environment and exchange data among devices [8]. In determining the sensing layer of an IoT, the main concerns are:

- Cost, size, resource, and energy consumption. The things might be equipped with sensing devices such as RFID tags, sensors, actuator, etc., which should be designed to minimize required resources as well as cost.
- Deployment. The IoT end-nodes (such as RFID reader, tags, sensors, etc.) can be deployed one-time, or in incremental or random ways depending on application requirements.
- Heterogeneity. A variety of things or hybrid networks make the IoT very heterogeneous.
- Communication. The IoT end-nodes should be designed able to communicate with each other.
- Networks. The IoT involves hybrid networks, such as WSNs, WMNs, and supervisory control and data acquisition (SCADA) systems. The security is an important concern in sensing layer. It is expected that IoT could be connected with industrial networks to provide users smart services. However, it may cause new concerns in devices controlling, such as who can input authentication credentials or decide whether an application should be trusted. The security model in IoT must be able to make its own judgements and decision about whether to accept a command or execute a task. At sensing layer, the devices are designed for low power consumption with constraints resources, which often have limited connectivity. The endless variety of IoT applications poses an equally wide variety of security challenges:
- Devices authentication;
- Trusted devices;
- Leveraging the security controls and availability of infrastructures in sensing layer; and
- In terms of software update, how the sensing devices receive software updates or security patches in a timely manner without impairing functional safety or incurring significant recertification costs every time a patch is rolled out.

In this layer, the security concerns can be classified into two main categories:

- The security requirements at IoT end-node: physically security protection, access control, authentication, non-repudiation, confidentiality, integrity, availability, and privacy; and
- The security requirements in sensing layer: confidentiality, data source authentication, device authentication, integrity, availability, and timeless, etc. It summarizes the potential security threats and security vulnerabilities at the IoT end-node. To secure devices in this layer before users are at risk, the following actions should be taken: implement security standards for IoT and ensure all devices are produced by meeting specific security standards; build trustworthy data sensing system and review the security of all devices/components; forensically identify and trace the source of users; software or firmware at IoT end-node should be securely designed.

12.4.2 NETWORK LAYER

The network layer connects all things in IoT and allows them to be aware of their surroundings. It is capable of aggregating data from existing IT infrastructures and then transmits to other layers, such as sensing layer, service layers, etc. The IoT connects a verity of different networks, which may cause a lot of difficulties on network problems, security problems, and communication problems. The deployment, management, and scheduling of networks are essential for the network layer in IoT. This enables devices to perform tasks collaboratively. In the networking layer, the following issues should be addressed:

- Network management technologies including the management for fixed, wireless, mobile networks;
- Network energy efficiency;
- Requirements of QoS;
- Technologies for mining and searching;
- Information confidentiality; and
- Security and privacy.

Among these issues, information confidentiality and human privacy security are critical because of its deployment, mobility, and complexity. The existing network security technologies can provide a basis for privacy and security protection in IoT, but more works still need to do. The security requirements in network layer involve:

- Overall security requirements, including confidentiality, integrity, privacy protection, authentication, group authentication, keys protection, availability, etc.
- Privacy leakage. Since some IoT devices physically located in untrusted places, which cause potential risks for attackers to physically find the privacy information such as user identification, etc.
- Communication security. It involves the integrity and confidentiality of signalling in IoT communications.

- Overconnected. The overconnected IoT may run the risk of losing control of the user. Two security concerns may be caused: DoS attack, the bandwidth required by signalling authentication can cause network congestion and further cause DoS; keys security, for the overconnected network, the keys operations could cause heavy network resources consumption.
- MITM attack. The attacker makes independent connections with the victims and relays messages between them, making them believe that they are talking directly to each other over a private connection, when in fact the attacker controls the entire conversation.
- Fake network message. Attackers could create fake signalling to isolate/ misoperate the devices from the IoT.

The network infrastructure and protocols developed for IoT are different with the existing IP network, special efforts are needed on following security concerns: authentication/ authorization, which involves vulnerabilities such as password, access control, etc.; secure transport encryption, it is crucial to encrypt the transmission in this layer.

12.4.3 Service Layer

In IoT, the service layer relies on middleware technology, which is an important enabler of services and applications. The service layer provides IoT a cost-effective platform where the hardware and software platforms could be reused. The IoT illustrates the activities required by the middle service specifications, which are undertaken by various standards developed by the service providers and organizations. The service layer is designed based on the common requirements of applications, application programming interfaces (APIs), and service protocols. The core set of services in this layer might include following components: event processing service, integration services, analytics services, UI services, and security and management services. The activities in service layer, such as information exchange, data processing, ontologies databases, communications between services, are conducted by following components:

- Service discovery. It finds infrastructure can provide the required service and information in an effective way.
- Service composition. It enables the combination and interaction among connected things. Discovery exploits the relationships of things to find the desired service, and service composition schedules or re-creates more suitable services to obtain the most reliable ones.
- Trustworthiness management. It aims at understanding how the trusted devices and information provided by other services.
- Service APIs. It provides the interactions between services required by users. In recent, a number of service layer solutions have been reported. The SOCRADES integration architecture is proposed that can be used to interact between applications and service layers effectively; things are abstracted as devices to provide services at low-levels as network discovery services,

metadata exchange services, and asynchronous publish and subscribe event, a representational state transfer is defined to increase interoperability between loosely coupled services and distributed applications. In the services layer a service provisioning process can provide the interaction between applications and services. It is important to design an effective security strategy to protect services against attacks in the service layer. The security requirements in the service layer include:

- Authorization, service authentication, group authentication, privacy protection, integrity, integrity, security of keys, non-repudiation, anti-replay, availability, etc.
- Privacy leakage. The main concern in this layer involves privacy leakage and malicious location tracking.
- Service abuses, in IoT the service abuse attack involves: illegal abuse of services; abuse of unsubscribed services.
- Node identify masquerade.
- DoS attack, denial of service.
- Replay attack, the attacker resend the data.
- Service information sniffer and manipulation.
- Repudiation in service layer, it includes the communication repudiation and services repudiation. The security solution should be able to protect the operations on this layer from potential threats. Ensuring the data in service layer secure is crucial but difficult. It involves fragmented, full of competing standards and proprietary solutions. The SoA is very helpful to improve the security of this layer, but the following challenges still need to be faced when building an IoT services or application: data transmission security between service and/or layers; secure services management, such as service identification, access control, services composite, etc.

12.4.4 APPLICATION-INTERFACE LAYER

The application-interface layer involves a variety of applications interfaces from RFID tag tracking to smart home, which are implemented by standard protocols as well as service-composition technologies. The requirements in application-interface layer strongly depend the applications. For the application maintenance, the following security requirements will be involved:

- Remote safe configuration, software downloading and updating, security patches, administrator authentication, unified security platform, etc. For the security requirements on communications between layers:
- Integrity and confidentiality for transmission between layers, cross-layer authentication and authorization, sensitive information isolation, etc.

In IoT designing the security solutions, the following rules will be helpful:

1. Since most constrained IoT end-nodes work in an unattended manner, the designer should pay more attention to the safety of these nodes;

2. Since IoT involves billions of clustering nodes, the security solutions should be designed based on energy efficiency schemes; and

3. The light security scheme at IoT end-nodes might be different with existing network security solutions; however, we should design security solutions in a big enough range for all parts in IoT. It summarizes the security threats and vulnerabilities in IoT application interface layer. We analyse the security threats and potential vulnerabilities in application-interface layer. The application-interface layer bridges the IoT system with user applications, which should be able to ensure that the interaction of IoT systems with other applications or users are legal and can be trusted.

12.4.5 CROSS-LAYER THREATS

Information in the IoT architecture might be shared among all of the four layers to achieve full interoperability between services and devices. It brings a number of security challenges such as trust guarantee, privacy of the users and their date, secure data sharing among layers, etc. In the IoT architecture described in Figure 12.2, information is exchanged between different layers, which may cause potential threats. The security requirements in this layer include security protection, securing to be ensured at design and execution time; privacy protection, personal information access within IoT system, privacy standards and enhancement technologies; trust has to be a part of IoT architecture and must be built in.

12.5 THREATS CAUSED IN MAINTENANCE OF IOT

The maintenance of IoT can cause security problems, such as in configuration of the network, security management, and application managements. It summarized the potential threats that can cause risks in IoT.

12.5.1 SECURITY AND PRIVACY ISSUES WITHIN THE INTERNET OF THINGS

IoT is expanding in the world rapidly. It is becoming more associated with the fabric of human daily lives. IoT is gradually becoming a vital component of human infrastructure. In this regard, the question of its security has become vital. The security of the IoT eco-system is based upon some principles such as confidentiality, availability, integrity, possession, utility and authority.

These ingredients were improved by adding some other ingredients such as reliability, robustness, safety, performability, resilience, and survivability. Some of the issues covering IoT applications subjected to security challenges are heterogeneity, authentication and identity management, authorization and access control, accountability, health issues, logistics, smart grid issues, and so on. Privacy issue appears to be a major concern so far as IoT is concerned. Through IoT activities, enormous quantities of data are exchanged. Concerned users' confidence about the preservation of their privacy is to be ensured.

If the users are safeguarded and satisfied, they would enhance their usage of IoT. Modern technologies such as virtual private networks, transport layer security, and private information retrieval have been developed to protect privacy. These are essential as IoT-related technologies have already entered the homes. Data associated with a fridge (smart fridge) in the home can be used to realize the food habits of the concerned residents. This is an essential information for life insurance companies to detect health conditions of residents. Hence, these vulnerabilities are to be regulated properly. In the UK, General Data Protection Regulation has been framed to address these situations. The IoT-related technologies have advanced a lot. Even toys fitted with appropriate sensors can expose identifiable information of children. This might be instrumental to locate the child. These toys may also be used to act a surveillance device. It is otherwise vulnerable. Hence, strict regulations should be framed to check this. Developing countries have already framed appropriate regulations to control these affairs. In India, however, this type of measure is absent in the IoT draft policy. Authorities should pay attention to this.

12.5.2 VARIOUS CHALLENGES IN IoT DEVELOPMENT

The various challenges in the development of IoT are discussed as follows.

Data management challenge: IoT-based systems gather a large amount of data from the heterogeneous environment that need to be processed and stored. All the data collected by the user from various networks may not be useful for future purpose and need not be stored for backup. Consequently, few organizations are trying to prioritize data for operations or backup based on needs and value. The present architecture of IoT is not efficient to deal with all the problems related to gathering and storage of data and hence need to be addressed.

Data mining challenge: As more and more data is available to the user for processing and analysis, data mining tools are becoming the need of the hour. Data mining allows enterprises to predict future trends by sorting through large data sets to identify the patterns and solve problems through data analysis. Researchers use various data mining approaches like multidimensional databases, machine learning, soft computing, etc. for processing their data. Therefore, it is becoming necessary to incorporate the use of data mining techniques in the field of IoT and also there is a need for more competent data analysts.

Security challenge: With the growing development in the field of IoT, more and more devices are getting connected to it. But each new device connected increases the security concerns surrounding IoT. Ten years ago, we had to worry about protecting our computers only. Five years ago, we had to worry about protecting our smartphones. Now we have to worry about protecting our car, our home appliances, our wearables and many other IoT devices. In 2014, a study by Hewlett-Packard revealed that 70 per cent of the most commonly used IoT devices contain serious vulnerabilities. The authors in [6] discussed the various existing security challenges – interoperability, resource constraints, privacy Internet of Things security 297 protection and scalability. Therefore, security in IoT is the biggest concern at present and needs the attention of researchers.

Privacy challenge: As IoT is dealing with a large amount of data, this creates more entry points for hackers and leaves sensitive information vulnerable. Solving IoT privacy problems has a long way to go. An unauthorized person can track the personal details of a user and can make the unnecessary changes to it. Thus, the issue of privacy in the IoT motivates researchers to design new threat and attacker models that can be applied to IoT architectures and design methods for ensuring the privacy of IoT applications and architectures.

Chaos challenge: IoT is in a period of chaos as there are too many standards, ecosystems to connect billions of devices and applications together. In a hyper-connected world, a single error can disrupt the working of the whole system. For example, medical monitoring systems consist of a large number of interconnected sensors, controller and communication devices. A controller may receive an incorrect signal, which may prove fatal to the patient. To prevent such chaos in the hyperactive connected world, we need to make every effort to reduce the complexity of connected systems and enhance the security and standardization of applications.

12.6 SECURITY ARCHITECTURE

Generally, the security architecture of IoT is divided into three layers as shown in Figure 12.2. The various layers are discussed as follows.

Perception layer: It is also known as the "Sensors" layer in IoT. It is divided into two parts: Perception node (data acquisition and data control) and Perception network (sends collected data to the gateway or to the controller). A key component is sensors for capturing and representing the physical world in the digital world. It collects, detects, and processes information and then transmits it to the network layer. It performs IoT node collaboration in local and short-range networks.

- Network layer: It serves the function of data routing and transmission to different IoT hubs and devices over the internet. Cloud computing platforms, internet gateways, switching and routing devices, etc. operate by using technologies, e.g. Wi-Fi, LTE, 3G and ZigBee. Network gateways serve as the mediator between different IoT nodes by aggregating, filtering, transmitting data to and from different sensors.
- Application layer: It is the topmost and terminal layer. It guarantees authenticity, integrity, confidentiality of data. The main feature of this layer is data sharing. The purpose of IoT or the creation of a smart environment is achieved on this layer. It provides personalized services according to the needs of users.
- Protocols and technologies used at each layer: This section describes the various protocols and technologies used at each layer of the security architecture of IoT.
- IEEE 802.15.4: It is the largest standard that defines the operation of low-rate wireless personal area networks (LR-WPANs). It is used at the perception layer, which defines frequency, power, modulation, and other wireless conditions of the link. The most frequently used frequency is a 2.4 GHz band and it uses carrier sense multiple access with collision avoidance for channel access. It defines two network topologies: star and peer-to-peer.

- Routing protocol for low power and lossy networks: It is a distance vector and source routing protocol that allows a sender to partially or completely specify the route the packet takes through the network. A destination-oriented directed acyclic graph (DODAG) represents the core of routing protocol for low power and lossy networks (RPL). A DAG is rooted at a single destination at a single DAG root (DODAG root) with no outgoing edges. RPL routers use two modes of operation:
  - Non-storing mode—routes messages toward lower level based on IP source routing; and
  - Storing mode—downward routing based on destination IPv6 addresses.

12.6.1 Security Principles

The main security principles for IoT include the following.

- Confidentiality: Confidentiality, in the context of information security, allows only authorized users to access sensitive and protected data. Specific mechanisms ensure confidentiality and safeguard data from harmful intruders. In IoT, it is very important to ensure that data is secure and is available to authorized users only. The data collected by the sensors must not be revealed to the neighboring nodes. Protection of data throughout the process of data collection and data aggregation is an important requirement.
- Integrity: Integrity is concerned with ensuring that data are real, accurate and safeguarded from unauthorized user modification. This feature can be imposed by maintaining end-to-end security in IoT. Data should come from the right sender and must be transmitted to the intended receiver. Also, the data should not be altered during the process of transmission.
- Availability: Availability is mainly concerned with the user accessing information or resources in a specified location and in the correct format. Anytime and anywhere availability of data is very important to the users of IoT. Devices and services must also be reachable and available when needed in a timely fashion.
- Authentication: Authentication in computer systems context is concerned with protecting the integrity of a message, validating the identity of the originator, non-repudiation of origin (dispute resolution). As many users are involved in the IoT and they need to interact with each other to exchange the information between them, it is necessary to have a mechanism to mutually authenticate entities in every interaction. Identification and authentication of other objects by a particular sensor or object are one of the main characterizations of IoT objects. The authors in [5] presented a comprehensive survey of authentication protocols for IoT based on the target environment and the various ways in which the authentication protocols for IoT may be improved.
- Lightweight solutions: Lightweight solutions aim at using the lesser resources of the devices. As IoT devices have inherent power and computational limitations, this kind of unique security feature was introduced in IoT. It is a restriction that must be considered while designing and implementing protocols. Device capabilities decide the type of algorithms is meant to be run on IoT devices.

- Heterogeneity: Heterogeneity is kind of an inherent characteristic feature of IoT as it aims at connecting device-to-device, human to device and human-to-human. IoT provides a connection between heterogeneous things and networks. Protocols must be designed to connect different entities with different capabilities, complexity, and vendors.
- Policies: Policies and standards must be designed to ensure that data will be managed, protected, and transmitted in an efficient way. The policies that are used currently in computer and network security are not applicable to IoT due to its heterogeneous and dynamic nature.
- Key management systems: There must be a lightweight key management system for all frameworks that can enable trust between different things and can distribute keys by consuming minimum resources.
- Non-repudiation: It is important to ensure that someone cannot deny the authenticity of their signature or message originated by them. A device cannot deny that it had not sent the previous message.
- Freshness/no reply: It is important to ensure that the data are recent and not that the old messages have been replaced. There are two types of freshness:
  - weak freshness—required by sensor measurements; and
  - strong freshness—used for time synchronization within the network.

12.6.2 OPEN QUESTIONS

With IoT evolving rapidly, the researchers, as well as experts from the industry, keep on asking questions regarding the various concerns of the IoT, ranging from the effective and efficient use of the IoT paradigm to security threats. Among these questions, approximately eight out of ten researchers/experts from the industry would definitely pick the privacy and security as the area of concern. One of the most common questions that definitely tops the charts is: "What is the biggest risk associated with the IoT?" and the answer most of the time is security. Thus, the security in the IoT should be a prime focus for most of the review papers being published in this field. In this chapter, we have tried to identify and ask what possibly can go wrong if a particular attack occurs at some particular level of the security protocol stack of IoT. Furthermore, how the possible solutions are advantageous or disadvantageous in terms of implementation in the future.

IoT is going to dictate terms in the field of research for a few upcoming decades for sure. Being a very promising field, it is quite challenging too. It will require the new and scalable architectures and protocols, which need to be quite the efficient ones. One of the prime aspects from the many challenging issues is security. The data exchanged between the different devices or applications are quite sensitive; thus, the security aspect plays a key role and needs to be addressed efficiently. Being heterogeneous in nature, IoT security architecture is vulnerable to different types of attacks at every layer. This indicates the urgent needs of developing general security policy and standards for IoT products. The efficient security architecture needs to be imposed but not at the cost of efficiency and scalability. We need to solve one issue keeping in mind the second one at the same time. We have recommended a detailed and extensive survey regarding the different types of attacks that can possibly attack at each layer of IoT security protocol stack.

One of the important things concluded in the chapter is that there are numerous kinds of attacks that affect the privacy and security of the private and critical data. Thus, countermeasures are required so that the privacy and security of the data and the data exchanging processes are ensured.

12.6.3 FUTURE DIRECTIONS OF RESEARCH

The future research directions mainly consist of dealing with the mitigation of the different security challenges at each layer of the security protocol of IoT. It is recommended to implement the upcoming proposals for IoT security using Wi-Fi Harlow with 6LOWPANIPSec in low equipment for security test. Also, the future cloud-based security services can improve the security of the IoT.

12.7 CONCLUSION

Research in the field of security in the IoT is still in the conceptual stage, which needs to be explored further to develop innovative, new security solutions and applications. In recent years, the research on this topic is very active, as the issue of security in IoT must be considered first during the development of IoT. In this chapter, we presented an extensive and comprehensive survey on the current state of the art in the IoT security along with the layered security stack of IoT. In addition, the various types of attacks that occur at the three layers of IoT security protocol stack – perception, networking and application layers – are elicited and explained in detail along with the possible solutions that can be immediately applied at each layer. Futhermore, the countermeasures at each layer along with the prospective advantages and disadvantages are proposed so that the researchers can get a clue in implementing the particular strategy.

BIBLIOGRAPHY

[1] P. C. Evans and M. Annunziata, Industrial internet: Pushing the boundaries of minds and machines, GE, Technical Report November 2012, pp. 1–37.
[2] T. H. Szymanski, "Supporting consumer services in a deterministic industrial internet core network", *IEEE Communications,* 54(6), 2016, pp. 110–117.
[3] G. Fettweis "The tactile internet", *ITU-T Technology Watch*, August 2014, pp. 1–24.
[4] M. Maier et al., "Tactile internet: Vision recent progress and open challenges", *IEEE Communications*, 54(5), 2016, pp. 138–145.
[5] T. H. Szymanski, "Securing the industrial-tactile Internet of Things with deterministic silicon photonic switches", *IEEE Access*, 4, 2016, pp. 8236–8249.
[6] T. H. Szymanski, "An ultra-low latency guaranteed rate internet for cloud services", *IEEE Transactions Networking*, 24(1), 2016, pp. 123–136.
[7] FIPS, *Announcing the Advanced Encryption Standard (AES),* Federal Information Processing Standards (FIPS), 2001, pp. 1–51.
[8] G. Varghese and N. Bjorner, "Network verification", ACM SIGCOMM tutorial, August 2015.
[9] E. Barker, *Guideline for Using Cryptographic Standards in the Federal Government: Cryptographic Mechanisms*, NIST special publication 800-175B, March 2016, pp. 1–79.

[10] A. Nordrum, "Quantum computer comes close to cracking RSA encryption", *IEEE Spectrum*, March 2016, available: bit.ly/1oSxb92.

[11] Y. A. Vlasov, "Silicon-CMOS integrated nano-photonics for computer and data communications beyond 100G", *IEEE Communications*, February. 2012, pp. 67–72.

[12] J. Borghoff et al., "PRINCE: A low-latency block cipher for pervasive computing applications", *Proceedings Internatiuonal Conference Theory and Application of Cryptology and Information Security*, 2012, pp. 208–225.

[13] A. Bogdanov et al., "PRESENT: An ultra-light-weight block cipher", *Proceedings International Workshop Cryptographic Hardware and Embedded Systems*, 2007, pp. 450–466.

[14] T. H. Szymanski, "Max-flow min-cost routing in a future internet with improved QoS guarantees", *IEEE Transactions Communications*, 61(4), 2013, pp. 1485–1497.

[15] M. Rezaee and T. H. Szymanski, "Demonstration of an FPGA controller for guaranteed-rate optical packet switching", *Proceedings IFIP/IEEE International Symposium Integrated Network Management (IM)*, Ottawa, Canada, 2015, pp. 1139–1140.

13 Anomaly Detection over SDN Using Machine Learning and Deep Learning for Securing Smart City

Reenu Batra, Virendra Kumar Shrivastava, and Amit Kumar Goel

CONTENTS

13.1 INTRODUCTION

Nowadays, machine learning has become one of the prominent applications of artificial intelligence. Many of the applications use a huge amount of data. Data analysis has become a major concern for all of these applications as it must be done in a smart and efficient manner. Different types of data are used for different applications. To discover the properties of data, perceive knowledge and learn from data, we use the concept of machine learning. Machine learning basically makes use of learning algorithms to design a model. In other words, we can define machine learning as

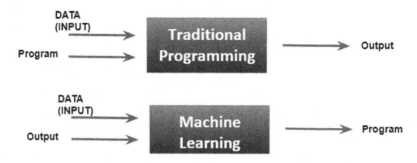

FIGURE 13.1 Machine learning

"learning from data." Data plays a vital role in the processing of machine learning techniques. Machine learning can be used in many applications like feature extraction, pattern recognition, speech recognition, email spam detection, data mining and classification etc. In other words, we can say machine learning is used in our day-to-day life. It allows computers to learn without being explicitly programmed. In traditional programming, in order to get the output, we need to input data as well as its logic. In contrast, machine learning (refer to Figure 13.1), both input and output, needs to be fed in the machine and it works in two different phases. In the first phase (training phase) we feed data (input and output) to the machine learning model and the model automatically creates its logic / learns from data and that can be evaluated in the second phase (testing phase).

13.2   SOFTWARE-DEFINED NETWORK

In order to deploy many applications, organizations make use of the network. If we talk about traditional networks, they are completely hardware-based. Traditional networks are basically static in nature. In order to run applications, connections are made by using physical devices such as routers and switches. A traditional network makes the use of a protocol for communication and ASIC (Application Specific Integrated Circuit) in order to implement the functionality of dedicated hardware devices. The layered architecture of a network basically consists of planes in operations. There are three planes in a network, namely control plane, data plane and management plane. One point to be noted is that the SDN control plane is a component of the router mainly used for routing table exchange and information, system configuration and management. All the data packets flowing over a network are handled by the data plane. In order to control and monitor all the hardware devices, the management plane is used. To achieve improvement in the working of traditional networks, SDN (Software Defined Networking) is used [1]. It mainly decouples the data plane and control plane of SDN. SDN is a networking approach that makes a networking framework agile and flexible. By making use of various software applications it makes network programmable, intelligent and centrally controlled. It mainly consists of different types of topologies to make it more flexible. It is an approach that makes use of open protocol like Open Flow (refer Figure 13.2).

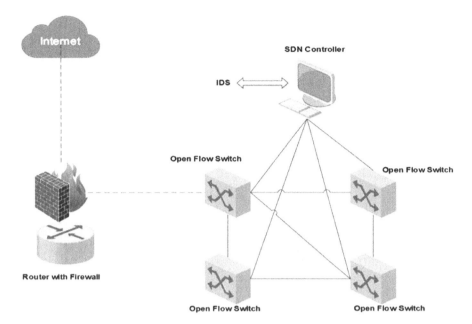

FIGURE 13.2   SDN architecture

SDN mainly separates control functions from forwarding functions to make a network. The architecture of SDN consists of three layers: application layer, control layer and infrastructure layer. The SDN controller resides on the control layer which can be synonymized as the "Brain of SDN." Applications and services of a network essentially work on the application layer. For many of the hardware devices like the router, switches reside on the infrastructure layer. Because of the agility and flexibility property of SDN, it can be used in various data-sensitive applications like big data. In SDN architecture, northbound API is used for communication between the SDN controller and the high component application. The communication between network devices (low-level components) and SDN controller is done with the help of the southbound API.

13.2.1   SECURITY OVER SDN

The security of data is a major concern in many networking applications. As the security of data increases, reliability also increases. Security over SDN can be achieved by protecting resources and information from unauthorized persons. Security issues rely upon SDN because of its dynamic nature, which may lead to alteration of data, data retention etc. If we talk about traffic over SDN [2], we can detect an attack if some deviation is found from normal profile behavior. The Intrusion Detection System (IDS) approach can be used in SDN to overcome the security challenge. IDS of SDN can be designed by using machine learning. SDN mainly makes use of OpenFlow protocol for communication [3]. In order to gain more efficiency in

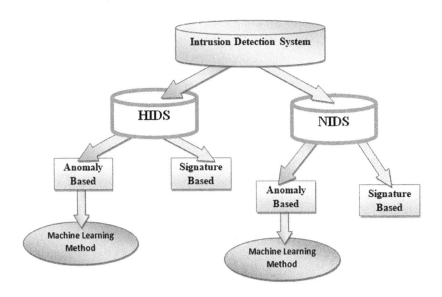

FIGURE 13.3   IDS, HIDS, and NIDS

attack detection, the deep learning approach can be used. Many other solutions can be applied on SDN for attaining security such as firewall, policy management and access control mechanism. But these solutions do not guarantee security over SDN. IDS are more promising compared to other security solutions. The IDS task is to monitor the whole network traffic and it detects unwanted traffic. IDS is mainly designed for those kinds of networks that are unable to handle huge traffic over the network and are not capable of detecting unwanted traffic or attacks. So security over SDN data traffic can be achieved by making use of IDS in order to detect different kinds of attacks. IDS can be grouped into two different categories (refer to Figure 13.3) based on its monitoring and analysis techniques used [4].

- Host-Based IDS (HIDS)
- Network-Based IDS (NIDS).

HIDS mainly runs on individual hosts in order to detect attacks in the whole computer system. The first intrusion detection system implemented is HIDS. In order to detect attacks on a host, it must be installed on that particular host. It primarily detects unofficial activities by comparing its current data files and logs with old data files. If it finds a mismatch, it flags as an anomaly [5]. On the other hand, NIDS detects all arriving in and leaving out traffic in a computer system in order to find unauthorized access or anomalous behavior of traffic. Beside HIDS and NIDS, there are further three types of IDS: Protocol-based Intrusion Detection System (PIDS), Application Protocol-Based Intrusion Detection System (APIDS) and Hybrid Intrusion Detection System (HIDS). PIDS is mainly used at the server for monitoring the protocol used

TABLE 13.1
Comparison of signature-based and anomaly-based IDS

Sr. No.	Parameter compared	Signature based	Anomaly based
1.	Implementation	Easy	Complicated
2.	Alarm rate	Low	High
3.	Reliability	High	Moderate
4.	Robustness	Low	High
5.	Scalability	Low	High
6.	Speed	High	Low

between the user and server. It mainly uses HTTPS protocol streams. APIDS basically works on a group of servers to monitor communication between various application-specific protocols. Detection methods in IDS fundamentally come in two categories.

The detection approach in IDS may be signature-based or it may be anomaly-based [7]. Signature-based IDS detects only known attacks. Various techniques can be used by signature-based IDS like pattern matching that make use of known attacks and patterns in order to identify attacks. As a signature of new attacks gets changed then signature-based IDS also needs to change [6]. In contrast, anomaly-based IDS is used for identification of unknown attacks. Unusual behavior of the network can be examined by using anomaly-based detection techniques. On behalf of different parameters, comparison of both signature and anomaly-based IDS can be done (refer to Table 13.1).

A graphical representation can be done of both signature- and anomaly-based IDS by taking three scale values: low, moderate and high (refer Figure 13.4).

13.2.2 Types of Network Attacks

An attack is a process or method that is executed by a hacker or attacker in order to violate network security. There are mainly four network attack groups (refer to Table 13.2) in which we can categorize different types of network attacks [8].

1. **Denial of Service (DoS) attack**: The attack class mainly consists of unwanted traffic over traffic, which mainly denies an authorized user to use its service. For example, teardrop, TCP SYN flooding.
2. **Probe attack**: The probe attack class mainly collects information about all network devices. PortSweep is essentially used to scan ports for analyzing services running on a particular host.
3. **User to Root (U2R) attack**: In this attack class, the attacker tries to gain super-user privileges. Buffer-flow comes in this category.
4. **Remote to Local (R2L) attack**: In this class, an attacker is capable of getting unauthorized access with the help of password sniffing. In order to gain access, attackers send packets over the remote system.

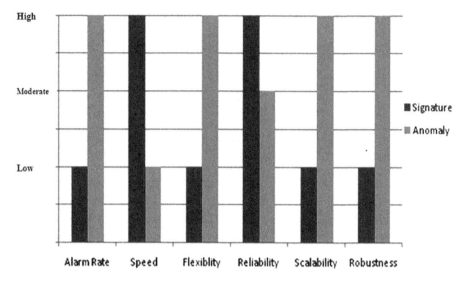

FIGURE 13.4   Graphical representation of signature-based and anomaly-based IDS

TABLE 13.2
Classes of attacks

Class 1	DoS	Denial of service
Class 2	U2R	User to root
Class 3	Probe	Attack related to the web structure
Class 4	R2L	Remote to local

13.2.3 EVALUATION METRICS

The input into IDS is network traffic and output will be either normal traffic or malicious traffic. IDS mainly classifies traffic as either normal or malicious. It uses a detection algorithm for classification of traffic [9]. Based on the output it sets off the alarm. The main feature of the anomaly detection algorithm is its detection rate. The feature or quality anomaly detection algorithm's detection rate can be measured by how accurately it detects attacks. This feature is known as accuracy. Another measured value of the detection algorithm is its recall and false alarm rate, Matthews correlation coefficient and F-measure. In IDS, we basically have a confusion matrix. A confusion matrix (refer to Table 13.3.) always consists of four values: T+, T-, F+, F-. Many of the measures can be calculated by making use of a confusion matrix [10].

1. **Accuracy:** It can be measured by the ratio of the total number of classifications that are performed correctly to the total number of all false (either positive or negative) and all true (either positive or negative) samples.

TABLE 13.3
Confusion matrix

T+	True positive	Total number of normal packets that actually predicted normal.
T-	True negative	Total number of attack packets that actually predicted attack.
F+	False positive	Total number of packets that are actually normal but declared as an attack
F-	False negative	Total number of packets that are actually attacked but declared as normal

$$Accuracy = (T+ + T-)/ (T+ + F+ + T- + F-)$$

2. **Precision:** It can be measured by ratio of true positive classification to all true positive and all false positive classification. It is inversely proportional to false alarm rate.

$$Precision = (T+)/ (T+ + F+)$$

3. **Recall**: It can be measured by the ratio of true positive classifications to all true positive and all false-negative classifications.

$$Recall = (T+)/(T+ + F-)$$

4. **F-measure**: Accuracy can be best measured with the help of F-measure.

$$F\text{-measure} = 2/(1/Precision + 1/Recall)$$

5. **Matthews correlation coefficient**: It takes the correlation coefficient of predicted and observed classifications. It gives the result between +1 and -1.

$$Matthews\ correlation\ coefficient = (T+ \times T- - F+ \times F-) / ((T+ + F+)$$
$$(T+ + F+) (T- + F+)(T- + F-))1/2$$

13.2.4 MACHINE LEARNING ALGORITHMS FOR ANOMALY DETECTION

Machine learning is used gradually in many of the running applications because of its capability of training a host or computer without any explicit programming. An advantage is that all the computers and controllers can work in a distributed environment and can work in a similar manner. By making use of machine learning algorithms, we can train a computer to find anomalies on the network as well. Many of the classification techniques are implemented by making use of ML [11]. Nowadays, many machine learning algorithms are used gradually in classification and regression.

Machine learning is a branch of artificial intelligence (AI). AI is a technique that mainly makes a machine intelligent and think like a human being. In order to train a robust version of AI, we use machine language. Machine learning is mainly based on statistical methods that train a machine learning model for predictions. In machine learning the whole concept is just gathering data, learning from data and finally making decisions based on the learning data. Machine learning algorithms give us high accuracy on the detection of the anomaly over SDN. The machine learning approach may be supervised, unsupervised and semi-supervised learning. In the case of supervised learning, we have input labeled data [12]. The model learns from that labeled data and based on learning it predicts the classes of unlabeled data. Support vector machine (SVM) is a well-known classification method, which comes under supervised learning. SVM can be performed on NIDS and is compatible with high precision data. The random forest (RF) algorithm is another algorithm that is used for supervised learning [13]. RF can easily operate over uneven data. In an unsupervised learning scheme, input data is unlabeled. So, in order to predict the undistributed data, machine learning models learn structure and presentations from the unlabeled data. Principal Component Analysis (PCA) and Self-Organizing Map (SOM) are examples of unsupervised algorithms [14]. PCA is mainly used to perform clustering of data and before clustering, it also uses a feature selection approach. The K-means algorithm may be used for clustering and anomaly detection. SOM is an artificial neural network (ANN) that can be used to overcome payload on NIDS. In the case of semi-supervised, some labeled data and a large amount of unlabeled data is used in training of algorithms [15]. Spectral Graph Transducer and Gaussian Fields are mainly two semi-supervised classification methods. MPCK-means is a clustering method (semi-supervised) used to improve anomaly detection [16].

13.2.5 DEEP LEARNING ALGORITHMS FOR ANOMALY DETECTION

By using deep learning, a machine learning model can learn structure and representation from unlabeled data with many levels. Deep learning is basically a subfield of ML, which is further a subset of AI [17]. It is based on ANN having a multilayer concept. Applications of deep learning may include object recognition, object detection, intrusion detection, etc. Learning algorithms based on deep learning may be supervised or may be unsupervised [18]. Examples of supervised deep learning algorithms are Convolutional Neural Network (CNN) and Auto-Encoder. The major application of CNN is computer vision [19]. On the other hand, Deep Belief Network (DBN) and Recurrent Neural Network (RNN) are algorithms that can be used as a training algorithm in both a supervised and unsupervised manner. RNN can be used in speech recognition [20].

13.3 MACHINE LEARNING AND INTERNET OF THINGS

Hardware technology and software technology is going to be developed more and more and day by day. With the increase in the number of technologies, the volume of data is going to increase [20]. The Internet of Things (IoT) is found to be the best solution, which is mainly an extension of the internet to connect a large number of physical devices together with a single technology. IoT characterized the big data concept that can be used to handle data with varying quality and of different modality. Many smart IoT applications can be introduced with the characterization of big data. Many ML algorithms can be applied to IoT data in order to get and process information. In IoT, there is basically the concept of wired communication, wireless communication, high speed internet, sensors and actuators. The IoT framework is mainly designed to perform activities with better performance and intelligently. As from the different devices, raw data can be collected and by processing of raw data we can extract the knowledge. In order to extract new data, IoT mainly uses the concept of data science. Data science mainly includes all the data mining activities, machine learning techniques and some other processing methods that can be applied to extract knowledge. In order to handle large IoT data, the big data concept is used. Big data mainly accomplishes high volume and high velocity data, and high variety data that must be processed with cost-effective and innovative data processing techniques [21].

From the challenges that came in to insight, the smart data concept was introduced. All the challenges related to volume of data, different variety of data and velocity of data can be threatened by smart data. Smart data technology mainly transforms raw data into smart data by using some methods that may lead to high productivity, effectiveness and efficiency. In short, IoT data is basically handled by smart data.

13.4 IOT AND SMART CITY

In order to develop the smarter environment, IoT plays a major role. Recently, many of the industries rely on IoT in order to overcome their expenses. Performance of devices can be enhanced if they all are connected in IoT environment so that these devices will be able to perform their functions and operations automatically without any user intervention. Basics consumption of resources like time, money and energy can be reduced by using a smart environment with IoT [21]. Sensors, interconnected network, data analyzing and system monitoring are major components of IoT. Most necessary and sufficient condition for IoT is connectivity of devices. Enhanced communication protocols are also required for effective IoT environment. In IoT there are mainly four types of communication protocols.

1. **Device-to-device protocol:** This device-to-device protocol mainly resides between two nearby devices like two cellular mobile phones.
2. **Device-to-server protocol**: This type of protocol is required to reside between a physical device and a server. The transfer of data mainly done between device to server or from server to device. Cloud processing is mainly known to be this type of protocol.
3. **Server-to-server protocol**: As many of servers reside over network for data handling. This type of protocol exists between two servers as servers are also responsible for transferring of data between each other. Cellular networking is base of this type of protocol networking.

In short, in IoT all the information is collected from the environment by making use of sensors. All sensors mainly collect data from the environment and then this data is processed, and further knowledge is extracted from the raw data. Then this knowledge is sent to other connected devices in the network.

The smart city concept also gains popularity in these recent trends. As with increase in population, problems of large-scale urbanization are also going to increase. These problems can be obsolete if we use the concept of IoT in cities [22]. All the city's assets can be connected , and smart monitoring devices can be used in order to see the live report of any part of city. All the tasks like traffic management, water management, energy management can be monitored, managed and further controlled by using the IoT smart city concept. We can consider the smart city with IoT as a use case because a city is an important part of society and all human life aspects are related to city. Smart energy, smart mobility and smart citizens are important aspects as they can have a direct impact on smart society.

13.5 SMART ENERGY, SMART MOBILITY, AND SMART CITIZENS

If we talk about smart energy, there must be less consumption of energy. Smart leak monitoring and smart energy measures are required to establish a smart city with smart energy. Having smart energy measures means a smart grid can be deployed in a smart city. Many hardware devices can be monitored by using a smart grid. It mainly requires managing time series data and further analyzes the data on demand. In order to optimize resource consumption, smart leak monitoring system can be introduced in a smart city. On the other hand, for a high-quality life, smart mobility also plays major role. Smart car introduction may also lead to more secure life as it can easily detect the movement of other cars on the road with help of sensors [22]. We can get information about shortest route of a destination by collecting data from movement of other car/vehicle movement [22]. That is possible only with help of wireless connections between devices and IoT sensors. The main advantage of having IoT sensors is that it may also lead to traffic control as much of the data can be collected with the help of road cameras and car sensors. In this way, accurate information about a particular route can also be collected. This will help in improvement in public transportation and will also lead to less wastage of time of passenger. Crime monitoring can also be done by using smart city concept. As with the help of sensors, we can detect the movement of citizens. All the citizens' records can be accessed any time so that unauthorized citizens will not be allowed to enter a smart city without verification.

13.6 BIG DATA AND IOT WITH MACHINE LEARNING

As in a smart IoT environment, data is generated in continuous manner. A major challenge in data collection from IoT-enabled devices is that the frequency of data delivered from devices may be mismatched. As a result, there may be loss of data. In order to cope with this challenge, different big data analytic methods can be applied. The dynamic nature of data also needs to be handled, especially in the smart city concept when the sensors generate different data according to different location and time [23]. Gaining more quality of information (QoI) is a major task as data is collected from different sources and so that data can be mismatch. QoI mainly depends upon a device's noise and precision of data collection. In order to get QoI, there high abstraction services are required. IoT data mainly have different features depending upon big data, data quality and data usage. Based upon big data, IoT data can be of different variety, different volume, and of different velocity. If we talk about data quality, IoT data can be data from a dynamic source, can handle redundant data, data of different granulate, etc. Based on data usage, IoT data is always complete and noiseless [23].

For smart data analysis, many of learning algorithms can be applied. A learning algorithm mainly takes input as a set of samples. These set of samples are known as a training data set. A learning algorithm can be supervised, unsupervised or reinforcement learning. In the case of supervised learning algorithms, in order to predict the unknown data, learning representations are done from the labeled input data. SVM, random forest and regression are the examples of supervised learning algorithms. SVM plays a major role in classification problems because of its practicability in computation and robustness in classification. The applicability of SVM is basically on high dimensional data. For ongoing training, SVM uses training instances.

On the other hand, the Navie Baye's model is a simplified model based on the Bayesian probability model. It is based on the work of Thomas Baye's. Many of the problems are based on Bayesian Classification, which calculates the probability of a hypothesis which claims the belonging of the data to a particular class [24]. For example, if a person is suffering with a disease, it can be determined by a test. The result of the test can be positive or negative. If the result is positive, it means the person has that particular disease, otherwise it is negative. On the other hand, this task can be represented by making use of two variables, infected and clean. The value of infected will be true if the person has the disease, otherwise it will be false. The value of variable clean will be true if the test performed gives a positive result otherwise it will be false.

In order to improve the efficiency of classification task, RF (random forest) works as ensemble classifier. Classification errors in RF are minimum compared to other classification algorithms. Many decision trees collectively form a random forest.

Whenever a node in a tree is to be split, some features need to be considered like size of node, number of trees and number of features used for splitting. One major advantage of RF is that all the generated forests can be used in future for references.

13.7   MACHINE LEARNING APPLICATIONS TO IOT AND IOT SECURITY

Routing control mechanisms can easily be implemented by collaboration of ML algorithms with IoT. By making use of ML with IoT at a single point we can easily get the best route information of the destination. For example, Google Maps and other GPS in car can be used to get destination information. If we talk about smart city living life, sensors like heat sensors, temperature sensors and humidity sensors can give minute-by-minute information about the day. This can be accomplished using the ML algorithm with IoT. Many of the government agencies, energy plants, water and gas plants, manufacturing companies, healthcare offices and retail stores can receive benefits from both ML and IoT.

IoT systems are more vulnerable to risk. There are many challenges that come in front when a thought arises for security of IoT systems. IoT systems are multipart models and used in many applications like smart cities, healthcare etc. A security system defined for one IoT application may not be compatible for another application [25]. A second thing about IoT systems is that they mainly use different platforms for different kind of devices. So it becomes a very challenging task to develop a security mechanism that will be applicable to all devices in an IoT system. IoT systems mainly deal with confidential data like biometric, fingerprints and travelling history etc. All of this personal information is stored on devices. Security is a major concern here as one of these devices may be turned in to eavesdropping device. There may be natural disaster conditions like earthquakes and floods that may lead to physical threat. In that case, security will become a major concern.

13.8 FEATURES OF IOT ENVIRONMENT

There are many features that mainly characterize an IoT system.

1. **Diverseness:** An IoT system mainly has different devices with different configuration, different communication protocols communicating on different platforms. This can be known as heterogeneous behavior of the IoT system and sometimes because of this heterogeneity attack threats may increase.
2. **Large-scale deployment**: IoT systems mainly have billions of interconnecting devices. All of these devices communicate to each other and send data over the internet.
3. **Local and global connectivity**: Connectivity among IoT-enabled devices may be local or global depending upon the application. For example, if we talk about the connecting devices in a car then it is basically local connection. In the case of data transfer over a smart home or smart city by mobile infrastructure, it is called global connection.
4. **Intelligence:** All the IoT devices operate in a timely manner intelligently. Decisions are made based on information processed on time.
5. **Point-to-point communication:** All devices in an IoT environment are directly connected to each other without intervention of central connectivity.
6. **Low power consumption**: In an IoT environment a number of devices connected over the internet may lead to low power consumption.
7. **Dynamic changes in network:** A number of devices in IoT work upon the requirement. Their state changes as per timeline. For example, sleep to wake and wake to sleep.

These are some major characteristics of IoT devices or IoT environment. These characteristics may be incorporated to make a smart city, which in future may lead to a smart society. For the soft functioning of all these devices, we must take care of these IoT devices and internet connection.

13.9 CONCLUSION

In this chapter, the concept of Intrusion Detection System (IDS) is explained. IDS may be implemented on a host or on a network. Intrusion can be detected on a machine by using an algorithm. Machine learning (ML) and deep learning (DL) are two subfields of artificial intelligence (AI) which may be used in detection of intrusion over Software Defined Networking (SDN). SDN is different from traditional networking as it decouples control plane from forwarding plane. Intrusion attacks may be classified in to four different classes: DoS attacks, Probe attacks, U2R attacks, R2L attacks. A machine learns by using any of ML or DL algorithms. The algorithm may be supervised, or it may be unsupervised. This chapter also explains how our ML can relate to IoT and what is the role of IoT in order to get smart environment. IoT device sensors plays major role in smart city traffic management, resource management and smart citizenship. IoT with big data characterization leads to many revolution tasks. Many ML algorithms like KNN, RF, decision tree etc. can be implemented in IoT. Besides this, IoT can be used in many of industries because of its many features. IoT security is also a major concern as it sends or receive confidential information.

REFERENCES

[1] McKeown, N., Anderson, T., Balakrishnan, H., et al. (2008) "OpenFlow: Enabling innovation in campus networks," *ACM SIGCOMM Computer Communication Review*, 38(2), pp. 69–74.

[2] Sen, S., Gupta, K. D., and Manjurul Ahsan, M. (2019) "Leveraging machine learning approach to setup software-defined network (SDN) controller rules during DDoS attack," *Algorithms for Intelligent Systems*, Singapore: Springer, Proceeding of International Joint Conference on Computational Intelligence, pp. 49–60.

[3] Jain, S., Kumar, A., Mandal, S., Ong, J., Poutievski, L., Singh, A., Venkata, S., Wanderer, J., Zhou, J., Zhu, M. et al. (2013) "B4: Experience with a globally-deployed software defined wan," *ACM SIGCOMM Computer Communication Review*, 43(4), pp. 3–14.

[4] Pervez, M. S., and Farid, D. M., "Feature selection and intrusion classification in NSL-KDD cup 99 dataset employing SVMs," *The 8th International Conference on Software, Knowledge, Information Management and Applications (SKIMA 2014)*, Dhaka: IEEE Xplore, 2014, pp. 1–6.

[5] Gude, N., Koponen, T., Pettit, J., Pfaff, B., Casado, M., McKeown, N., and Shenker, S. (2008) "Nox: Towards an operating system for networks," *ACM SIGCOMM Computer Communication Review*, 38(3), pp. 105–110.

[6] Jadidi, Z., Muthukkumarasamy, V., and Sithirasenan, E. (2015). "Performance of flow-based anomaly detection in sampled traffic," *IEEE, 9th International Conference on Signal Processing and Communication Systems (ICSPCS)*, 10, pp. 512–520.

[7] Winter, P., Hermann, E., and Zeilinger, M. (2011) "Inductive intrusion detection in flow-based network data using one-class support vector machines," 4th IFIP international conference on new technologies, mobility and security (NTMS), IEEE, pp. 1–5.

[8] Mehdi, S. A., Khalid, J., and Khayam, S. A. (2011) "Revisiting traffic anomaly detection using software-defined networking," in *International Workshop on Recent Advances in Intrusion Detection*. Cham: Springer, pp. 161–180.

[9] Braga, R., Mota, E., and Passito, A. (2010) "Lightweight ddos flooding attack detection using nox/openflow," in *IEEE 35th Conference on Local Computer Networks (LCN)*. IEEE, pp. 408–415.

[10] Kokila, R., Selvi, S. T., and Govindarajan, K. (2014) "DDos detection and analysis in SDN based environment using support vector machine classifier" in *Sixth International Conference on Advanced Computing (ICoAC)*, IEEE, pp. 205–210.

[11] Phan, T. V., Van Toan, T., Van Tuyen, D., Huong T. T., and Thanh, N. H. (2016) "Openflowsia: an optimized protection scheme for software-defined networks from

flooding attacks," in *IEEE Sixth International Conference on Communications and Electronics (ICCE)*, IEEE, pp. 13–18.

[12] Tang, T., Mhamdi, L., McLernon, D., Zaidi, S. A. R., and Ghogho, M. (2016) "Deep learning approach for network intrusion detection in software-defined networking," in International conference on wireless networks and mobile communications, WINCOM16, Fez, Morocco.

[13] Louridas, P., and Ebert. C. (2016) "The immense and positive implications of machine learning for humanity," *IEEE Software*, 33(5), pp. 110–115.

[14] Meng, Y. X. (2011) "The practice on using machine learning for network anomaly intrusion detection" in *International Conference on Machine Learning and Cybernetics (ICMLC)*, vol. 2. IEEE, pp 576–581.

[15] Tavallaee, M., Bagheri, E., Lu, W., and Ghorbani, A.-A. (2009) "A detailed analysis of the KDD cup 99 dataset," in Proceedings of the Second IEEE Symposium on Computational Intelligence for Security and Defence Applications.

[16] WEKA, "Data Mining Machine Learning Software." www.cs.waikato. ac.nz/ml/wcka,2018.

[17] Ingre, B., and Yadav, A. (2015) "Performance analysis of NSL-KDD dataset using ANN," in *2015 International Conference on Signal Processing and Communication Engineering Systems*, Guntur: IEEE Xplore, pp. 92–96.

[18] Abubakar, A., and Pranggono, B. (2017) "Machine learning based intrusion detection system for software defined networks," in Proceedings of the 2017 Eighth International Conference on Emerging Security Technologies (EST), IEEE.

[19] Ashraf, J., and Latif, S. (2014) "Handling intrusion and DDoS attacks in Software Defined Networks using machine learning techniques," in *National Software Engineering Conference*, Rawalpindi: IEEE Xplore, pp. 55–60.

[20] Meryem, A., Samira, D., Bouabid, El O., and Lemoudden, Mouad (2017) "A novel approach in detecting intrusions using NSLKDD database and Map Reduce programming," in *14th International Conference on Mobile Systems and Pervasive Computing (MobiSPC2017), ScienceDirect* vol. 110, Science Direct, pp. 230–235.

[21] Dhanabal, L., and Shantharajah, P. (2015) "A study on NSL-KDD dataset for intrusion detection system based on classification algorithms," *International Journal of Advanced Research Computer Communications Engineering*, 4(6), pp. 446–452.

[22] Saeid, M., Rezvanab, M., Barekatainc, M., Adibia, P., Barnaghid, P., and Shethb, A. P. (2018) "Machine learning for internet of things data analysis: A survey," *Digital Communication and Networks* 4(3), pp. 161–175.

[23] Hussain, F. , Hussain, R., Hassan, S. A., and Hossain, E. (2020) *Machine Learning in IoT Security: Current Solutions and Future Challenges*, IEEE Communications Surveys and Tutorials. doi: 10.1109/COMST.2020.2986444.

[24] Al-Garadi, M. A., Mohamed, A., Al-Ali, A., Du, X., Ali, I., and Guizani, M. (2020) "A survey of machine and deep learning methods for Internet of Things (IoT) security," IEEE Communications Surveys and Tutorials. *Computer Science.* doi: 10.1109/COMST.2020.2988293.

[25] Yan, Z., Zhang, P., and Vasilakos, A. V. (2014) "A survey on trust management for Internet of Things," *Journal of Network and Computer Applications*, 42, pp. 120–134.

14 Edge AI with Wearable IoT

A Review on Leveraging Edge Intelligence in Wearables for Smart Healthcare

Nidhi Chawla and Surjeet Dalal

CONTENTS

Internet of Things (IoT) and artificial intelligence are currently two of the most powerful and trending pieces for innovation in healthcare. The combination of IoT and artificial intelligence is a boon to the healthcare industry because both technologies are significantly improving the quality of life and making everything more personalized and intelligent. Recently, wearable devices such as smartwatches and fitness trackers have rapidly emerged as ubiquitous technology of the IoT and popularly known as wearable IoT (WIoT). Due to their sensing capability, individuals are seamlessly tracked by wearable sensors for personalized health and wellness information—body vital parameters, physical activity, behaviors, and other critical parameters. With recent advancements, future generations of WIoT are going to transform the healthcare sector by involving AI technology in order to improve functionalities and provide users with real-time insights and statistics along with assistance for better lifestyle choices. This chapter begins with an introduction of leveraging artificial intelligence in Internet of Things that covers various roles and the significance of different technology like WIoT, edge computing and edge intelligence in smart healthcare. This chapter aims to provide a comprehensive review of the current state of the art at the intersection of deep learning with edge computing in order to provide edge intelligence in wearable devices.

14.1 INTRODUCTION

One of the most critical needs today is access to good healthcare. Technology in the last few years has enabled us to closely monitor our health. Rising healthcare costs, population aging and the need for medical diagnosis are driving the sensors market in medical applications. Currently, smartphones, wearable devices and IoT are becoming more affordable and ubiquitous [45].

A recent research and markets report predicted that the IoT healthcare market is expected to grow from US$41.22 billion in 2017 to US$158.07 billion by 2022, at a compound annual growth rate of 30.8% demonstrating how medical care and healthcare services represent one of the most attractive fields for the development of IoT [54]. Many commercial products, such as the Apple Watch, Fitbit, and Microsoft Band and smartphone apps including Runkeeper and Strava, are already available for continuous collection of physiological data. These products typically contain sensors that enable them to sense the environment, have modest computing resources for data processing and transfer, and can be placed in a pocket or purse, worn on the body, or installed at home [34]. Wearable IoT today can provide doctors with the kind of patient data they never had before, enabling them to proactively diagnose patients before they head to a hospital for treatment. AI, meanwhile, has been helping to process the tremendous amounts of data, from scans to treatment reports, to identify disease patterns. This enables more accurate diagnosis that leads to the right treatment [55].

When it comes to combining AI and IoT in healthcare, chances are together they will improve operational efficiency in this field. Tracking (collecting), monitoring (analyzing), control, optimization (training), and automation (modelling, predicting)—these are the key steps that enable the smart and efficient application of AI algorithms in IoT devices. So that's why healthcare providers and device makers are integrating AI and IoT to create advanced medical applications and devices that can provide person-centric care for individuals, from initial diagnosis to on-going treatment options, while solving a variety of problems for patients, hospitals, and the healthcare industry. These AI-enabled wearable IoT devices will make healthcare treatments more proactive rather than preventive [56].

IoT applications generate enormous amounts of data by IoT sensors. However, sending all the data to the cloud will require prohibitively high network bandwidth. Recent research efforts are investigating how to effectively exploit capabilities at the edge of networks in order to support IoT and its requirements [21]. In edge computing, the massive data generated by different types of IoT devices can be processed at the network edge instead of transmitting them to the centralized cloud infrastructure owing to bandwidth and energy consumption concerns. It can provide services with faster response and greater quality in comparison with cloud computing (Yuan Ai et al.). Moreover, by enabling edge computing, crucial data can be transmitted from the ambulance to the hospital in real time, saving time and arming emergency department teams with the knowledge they need to save lives [27].

Billions of data bytes, generated at the network edge, put massive demands on data processing and structural optimization. Thus, there exists a strong demand to integrate edge computing and AI. So, the marriage of edge computing and AI has given rise to a new research area, namely, "edge intelligence (EI)" or "edge AI [3]. The fusion of AI and edge computing is natural since there is a clear intersection between them. Specifically, edge computing aims at coordinating a multitude of collaborative edge devices and servers to process the generated data in proximity, and AI strives for simulating intelligent human behavior in devices/machines by learning from data. However with edge AI, AI processing is now moving part of the AI workflow to a device and keeping data constrained to a device [1, 3].

This chapter highlights the recent advancements and associated challenges in employing wearable internet of things (WIoT) and body sensor networks (BSNs) for healthcare applications. In this chapter, there is detailed description about the application of AI in the modern healthcare system and the various algorithms used for implementing AI. The main focus of this chapter is to leverage edge intelligence in WIoT for smart healthcare. So, in this regard, a comprehensive smart healthcare framework is presented that will combine various emerging key technologies for a deep learning model toward training/inference at the network edge. Furthermore, it also provides the necessary background for potential future research initiatives in edge intelligence and recent breakthroughs in the application of AI with edge computing in healthcare.

This chapter is organized as follows: Section 14.2 includes work related to wearable technology, edge computing, deep learning and edge intelligence. Section 14.3 focuses on the various roles of AI, edge computing and WIoT for smart healthcare. Finally, conclusions are summarized in Section 14.4.

14.2 LITERATURE SURVEY

14.2.1 RELATED WORK FOR WEARABLE TECHNOLOGY

Baker et al. have proposed a generic model that can be applied to all IoT-based healthcare systems. The authors have discussed several wearable and non-intrusive sensors with particular focus on monitoring vital signs, blood pressure, and blood oxygen levels. Short-range and long-range communications standards are also compared in terms of suitability for healthcare applications [18].

Guk *et al.* have reviewed the latest advances in wearable sensor technologies to identify important biomarkers for noninvasive and possibly continuous monitoring of key diagnostic indicators. This chapter focus on the technical barriers and challenges in the development of wearable devices, and discusses future prospects on wearable biosensors for prevention, personalized medicine, and real-time health monitoring. One such technical challenge is the personal calibration of wearable devices whereas another challenge is the misalignment of wearables, which affects the quality and accuracy of the measurements [6].

Dias *et al.* has discussed important aspects in the wearable health devices area, listing the state-of-the-art of wearable vital signs sensing technologies and their system architectures and specifications. The authors have focused on traditional vital signs that have major importance and recent scientific developments on the area, such as electrocardiogram, heart rate, blood pressure, respiration rate, blood oxygen saturation, blood glucose, skin perspiration, capnography, body temperature, motion evaluation, cardiac implantable devices and ambient parameters [13].

14.2.2 RELATED WORK FOR EDGE COMPUTING

One way of addressing the resource constraints in IoT devices in terms of computation power and energy consumption is through offloading the computation to the gateway layer, often called edge or fog computing [8]. This approach can help improve performance and energy consumption to enable IoT devices to deliver real-time services. Several systems have adopted this approach by implementing IoT gateways employing resource-constrained devices (e.g., Raspberry Pi, Intel Edison) or personal mobile devices (e.g., smartphone, smartwatch) and demonstrated a good performance by reducing the amount of data to transfer to the cloud, and latency in applications [11].

Takiddeen *et al.* have conceptualized how smartwatches can be effectively utilized as edge devices within future IoT scenarios. In this regard, context computation, edge analytics, and computation off-loading are used to reduce power consumption for smartwatches [29].

Cooney *et al.* have developed a low-cost, portable, and wearable device, which can collect human activity data in real time and transfer it to a local server for computing and analysis. The proposed WIoT architecture-based device is suitable for rehabilitation in a home-based environment without the supervision of trained physiotherapists. Considering its cost-effectiveness, the design could also be used to provide efficient healthcare services to the people of low- and medium-income countries (LMICs). Such systems could be highly beneficial for LMICs as they have a burgeoning population but a shortage of staff [12].

14.2.3 RELATED WORK FOR DEEP LEARNING

At present, various studies have focused on compressing large-scale depth models to reduce computational complexity, reduce power consumption and memory requirements of devices. More and more researchers are committed to migrating deep learning technology to edge devices for IoT data analysis, in order to obtain real-time, efficient and accurate services.

As deep learning technology advances, many researchers apply it to human behavior recognition technology Examples of applications in the field of healthcare include detecting walking, running, and sitting in human activities through machine learning algorithms. Kwon *et al.* has identified active and static behavior in daily life behavior by using deep learning with streaming data obtained through a smartwatch. The overall accuracy of each model is 99.24% and 96.67% [31].

Uddin *et al.* have developed a multimodal human activity prediction system that is based on multiple wearable healthcare sensors (i.e., ECG, magnetometer, accelerometer, and gyroscope). In this, an RNN deep learning model is deployed on edge devices (e.g., laptop) for 12 different human activities. Using the proposed approach, a maximum 99.69% mean prediction performance has been achieved on the public dataset [9].

Fekri *et al.* have proposed pseudo code for a deep learning convolution neural network algorithm in order to detect 22 activities on real-life data coming from sensors in real-life smart devices. In this research, the authors have taken advantage of sensors that are already placed in smart devices like smartwatches or smartphones rather than using multiple sensors attached to the human body. In this algorithm, the author have defined a dynamic parameter as the window size and it will be varied with different activities. In future works it can be measured automatically based on training data and determining the repeated behavior in the dataset [32].

Zebin *et al.* have normalized the deep LSTM recurrent network on a smartphone for human activity recognition. For this, 32-bit floating-point precision weights are reduced to an 8-bits unsigned integer, which is known as quantization. It has been observed that almost all the size and execution time reduction in the optimized model are mostly due to weight quantization [14].

14.2.4 RELATED WORK FOR EDGE INTELLIGENCE

Edge intelligence, currently in its early stages, is attracting more researchers and companies from all over the world. To disseminate the recent advances of edge intelligence, Zhou et al. have conducted a comprehensive and concrete survey of the recent research efforts on edge intelligence. The authors have surveyed the architectures, enabling technologies, systems, and frameworks from the perspective of AI models' training and inference. However, the material in edge intelligence spans an immense and diverse spectrum of literature, in origin and in nature, which is not fully covered by this survey [3].

Wang *et al.* discussed potential enabling hardware of edge intelligence, i.e., customized AI chips and commodities for both end devices and edge nodes. Emerged edge AI hardware can be classified into three categories according to their technical architecture: (i) Graphics Processing Unit (GPU)-based hardware, which tends to have good compatibility and performance, but generally consume more energy, e.g., NVIDIA GPUs based on Turing architecture. (ii) Field Programmable Gate Array (FPGA)-based hardware, which is energy-saving and requires less computation resources, but with worse compatibility and limited programming capability compared to GPUs. (iii) Application Specific Integrated Circuit (ASIC)-based hardware, such as Google's TPU and HiSilicon's Ascend series usually with a custom design that is more stable in terms of performance and power consumption [1].

Yu *et al.* have proposed a hybrid architecture named EdgeCNN, which is deployed on smart devices that act as edge devices, giving real-time diagnostic capabilities. The authors have presented a streamlined and efficient model to identify ECGs with atrial fibrillation characteristics by using convolutional neural networks (CNN). This model can significantly reduce diagnosis latency and network I/O, and ease the pressure on the cloud platform. It has also significant advantages in delay, network I/O, convenience, and cost compared to the architecture based on the cloud platform only. In this, the author has protected the privacy of user data [15].

14.3   ROLE OF AI, EDGE COMPUTING AND WEARABLE IOT

14.3.1   Wearable IoT Framework for Smart Healthcare

Wearable IoT (WIoT) can be defined as a technological infrastructure that interconnects wearable sensors to enable monitoring human factors including health, wellness, behaviors and other data useful in enhancing individuals' everyday quality of life. The increased use of Internet of Things (IoT) technology in innovative electronic wearables has witnessed attractive and positive growth [45].

It is no longer sufficient to design standalone wearable devices, but it has become vital to create a WIoT ecosystem in which body-worn sensors seamlessly synchronize data to the cloud services through the IoT infrastructure. The following describes different components of WIoT architecture and their interconnections, a system that would benefit the healthcare industry in various ways [51].

Wearable Body Area Sensors (WBAS): WBAS are the front-end components of WIoT and unobtrusively envelop the body to capture health-centric data, primarily responsible for:

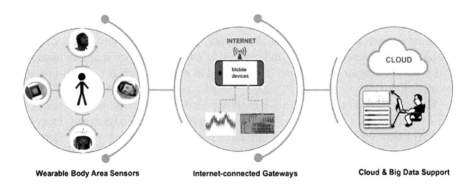

FIGURE 14.1   Architectural elements of WIoT

- Data collection—Directly from the body through contact sensors or from peripheral sensors
- Data preparation—Preparing data for on-board analysis for close-loop feedback/remote transmission.

Internet Connected Gateways: WBAS are rarely standalone systems due to their limited computing power and communication bandwidth and thus they need to transmit data to potent computing resources that are either companion devices (smartphones, tablets, and PCs or remotely located cloud computing servers). Companion devices are used as gateway devices that enable the information to flow from the sensors to the cloud or server centers for storage and further analysis using short-range communication technology (Bluetooth / NFC) used to exchange data with wearable sensors, and of heterogeneous networks, (WIFI and GSM) used to send the data to the cloud [45,51].

Cloud and Big Data Store: A cloud computing infrastructure can facilitate the management of wearable data and can support advanced functionalities of data mining, machine learning, and analytics.

- Routing protocols that can network smartphones and wearable sensors for handshaking and seamless data transfer,
- Event-based processing—Reduce unwanted data processing on resource constraint wearable sensors,
- Annotated data logs that can add activity-level information on top of clinical data for enhancing the accuracy of machine learning algorithms on the cloud,
- Person-centered databases that can store the personalized data of patients securely for longitudinal analysis, and
- Data visualization that can channelize the data to end-users such as physicians and patients to provide decision support and patient-physician interactions.

WIoTs can play a key role in personalizing treatment. Each disease manifests a unique set of symptoms that differ in intensity and in their spatiotemporal pattern based on the patient. Medication adherence monitoring is another aspect that WIoT can help patients with. Future research can focus on health pattern identification, where the algorithms in the wearable sensors and the gateways can detect a health aberration or an impending emergency. This is possible as the WIoT are in the vicinity of the patient [45].

14.3.2 Wearable IoT Devices

Wearable technology has recently become very popular, flooding the IT and fashion market with devices that can be worn or included in dresses and accessories. Their success is not just based on a fashion trend; rather, smartwatches, glasses, and bracelets offer the user a lot of useful functionalities as well as the possibility to collect and elaborate a vast plethora of information regarding the owner (i.e., number

TABLE 14.1
Vital physical signs measured by smart wearable systems

Vital physical signs	Sensors	Detection of symptoms
Electrocardiograph (ECG)	Skin electrodes	Heart rate, heart rate variability
Electroencephalogram (EEG)	Scalp placed electrodes	Brain activity
Electromyography (EMG)	Skin electrodes	Electrical activity of muscles
Blood Pressure (BP)	Cough pressure sensor	Status of cardiovascular system, Hypertension
Blood Glucose (BG)	Glucose meter	Amount of glucose in blood
Galvanic Skin Perspiration	Woven metal electrodes	Skin electrical conductivity
Respiration Rate(RR)	Piezoelectric sensor	Breathing rate, Physical activity
Body Temperature (BT)	Temperature Probe	Skin temperature, health state
Activity, Mobility and Fall	Accelerometer	Body posture, limb movement

of steps, hearth rate, level of stress, GPS position, etc.) and the environment (i.e., sound, light, altitude, etc.) [20].

A smart wearable system supported by technologies, such as wireless sensor networks and electronic care surveillance devices, can be implemented by using sensors, actuators, and smart fabrics for health evaluation and decision support. Current applications of smart wearable systems mainly work in monitoring vital signs, body movement, location, and fall prevention [13, 46]. Table 14.1 shows the important physical signs that can be measured though smart wearable systems.

The main advantage of using wearable devices instead of smartphones to support these applications is represented by their ability to stay naturally on the owner's body without requiring attention and handling: e.g., wearables do not need to be held in your hand, a position control sensor for Alzheimer patients could be hidden in a bracelet, etc. Indeed, the widespread popularity of wearable technology can lead to the generation of a huge amount of data that must then be efficiently gathered, stored, and further elaborated, thus requiring innovative solutions. Nowadays, wearable devices with the integration of AI and machine learning will lead the way of future healthcare. Just like smartwatches, continuous glucose monitoring devices, smart bandages, smart pills and remote patient monitoring—to name a few—will be common applications in the future healthcare industry [47]. Table 14.2 represents different types of wearable devices used for healthcare.

14.3.3 EDGE COMPUTING FOR SMART HEALTHCARE

To handle the huge data from the devices and sensors, the data are transferred to clouds for high-performance computing and vast data storage. Such combination of cloud and IoT (Cloud–IoT) can make the healthcare system eligible for real-time applications. However, it is still a big challenge to handle the massive data obtained from increasingly healthcare devices and sensors [18].

Furthermore, the execution of decisions made through the cloud computing platform may fluctuate due to network quality (such as latency, throughput, network I/O,

TABLE 14.2
Different types of wearable devices

Wearable IoT	Different types	Types of sensors	Activities	AI + WIoT applications
Smartwatches	Apple Watch, Samsung Galaxy Gear, Fossil Sports	Electrical heart sensor, optical heart sensor, gyroscope, light sensor, accelerometer, magnetometer, barometric pressure sensor, temperature sensor, oxymetry sensor	Heart rate monitoring, Exercise and activity, workout tracking (running, walking, cycling, swimming, yoga) mensuration tracking	1. Detect cardiac arrest, atrial fibrillation 2. Fall detection 3. Automatic workout detection 4. Gesture pattern recognition
Fitness tracker	JawBoneUP, Fitbit, Garmin	Accelerometer, light sensor, barometer, bioimpedance sensor, compass, altimeter, capacitive sensor, gyroscope, temperature sensor, optical heart sensor	Monitor and tracking fitness related metrices such as track steps, distance travelled, caloric intake, sleep intake and even heart rate and blood pressure	1. Heart rate variability 2. Auto exercise recognition 3. Automatic movement detection 4. Sleep detection

(continued)

TABLE 14.2   Continued
Different types of wearable devices

Wearable IoT	Different types	Types of sensors	Activities	AI + WIoT applications
Smart eyewear	Google Glass, Gear VR, Immerse VR	Image *sensors*, accelerometer, gyroscope, magnetometer, proximity sensor, IPD sensor	Recording patient session, video recording of surgery, patient diagnostic images	1. Real time facial and emotion detection 2. Help in autism spectrum disorder(ASD) 3. Voice recognition
Smart clothing	Haxoskin Smart shirt, Hi-tech iTBra, Sensoria Smart Sock	Temperature sensors, heart rate sensors, respiration sensors, activity sensors, pressure sensors	ECG and heartbeat monitor, HRV (allowing stress monitoring, effort, load and fatigue assessments), QRS events, and Heart rate recovery, breathing rate (RPM), sleep tracking, Pace, cadence	1. Early prediction of lung disease 2. Early breast cancer prediction

	Devices	Sensors	Functions	Applications
Smart jewelry	Bellabeat Leaf Urban, NFC Ring, Ringly Luxe Smart Bracelet, Fitbit Flex 2, Joule Earring Backing	Optical pulse sensors, 3D accelerometer, gyroscope and body temperature sensors.	Stress sensibility, menstrual cycle tracking, meditation and breathing exercises, alarm, break reminder, track basic activities like steps, distance, burned calories, continuous heart rate tracking	1. Detect sleep abnormality 2. Predict mood and stress levels
Smart footwear	Nike Adapt BB, Sensoria Smart Running Shoe, Under Armour Hovr	Accelerometer, gyroscope and pressure sensor, motion sensors	Monitor speed, distance travelled, foot strike, ground contact time and cadence.	1. Gait assistance 2. Obstacles detection 3. Blind navigation 4. Reduce diabetic amputations
Smart hearables	Vinci—smart headphone	Accelerometer, gyroscope, proximity sensor, optical heart rate sensors	Fitness tracking, heart rate monitoring	1. Near field voice recognition for emergency services 2. Gesture recognition

etc.), resulting in unpredictable response time between the cloud and IoT devices. As user requests increase, large amounts of data may flood the cloud platform at the same time and network I/O will face enormous challenges that will cause network data congestion and user requests to queue up. The data generated by IoT devices is continuous and uninterrupted. Once congestion occurs within a short period of time, the number of pending requests and data that are congested will increase and the response delay will be extremely high. It even leads to a serious crash of the entire system [27].

To lessen the pressure on the clouds, Edge-of-Things (EoT) computing has recently been proposed to act in-between the sensors and cloud. This paradigm is referred to as edge computing or edge analysis. "Edge" is any computing or network resources present between the source of the data and the cloud. For example, a smartphone is the edge between body sensors and the cloud. The edge devices basically offer computing and storage capabilities on a small scale in real time. They also increase efficiency, since only meaningful and already processed data will be sent to the centralized cloud for storage and further processing if required. So, edge analysis greatly improves results by eliminating data transfer latency from source to cloud for processing and storage [12][4][8].

Nowadays, edge computing is becoming increasingly important for health IT infrastructure for local and remote patient monitoring and telemedicine. Edge computing with intelligent IoT devices will be able to gather on-patient data, send the findings to their local clinic or surgery, and give almost real-time information to medical staff. Patient data could, potentially, be reviewed even if the patient is not present and has not made an appointment. This will help deal more proactively and efficiently with ongoing, long-term conditions, such as diabetes and cardiovascular diseases. It could also help with more general care for the elderly and for those with dementia. Further into the future, edge computing could help hospitals and healthcare trusts run a network of telemedicine booths that provide online, onscreen, real-time access to doctors and healthcare professionals [59].

As shown in Figure 14.2, Patients with wearable IoT medical devices can be diagnosed quickly and effectively on-site, and the information gathered from them can be fed back into the central servers whenever connections are re-established. By interfacing with an edge data center, IoT healthcare devices can extend the reach of existing networks, enabling medical personnel to access critical patient data even in areas with poor connectivity. A wearable health monitor is an example of a basic edge solution. It can locally analyze data like heart rate or sleep patterns and provide recommendations without a frequent need to connect to the cloud [58].

14.3.4 Edge Computing Framework

The system architecture as shown in Figure 14.3, stretches from the data sources located on or around patients to the service providers. It contains the following major components:

Hybrid Sensing Sources: A combination of sensing devices attached/near to the patients represents the set of data sources. Examples include body area sensor networks (including implantable or wearable medical and non-medical sensors),

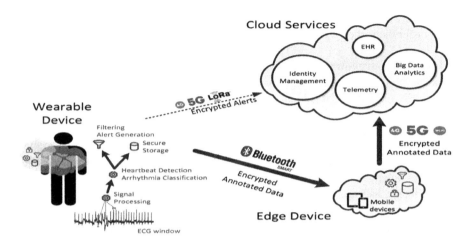

FIGURE 14.2 Integrating edge computing with wearable devices in healthcare

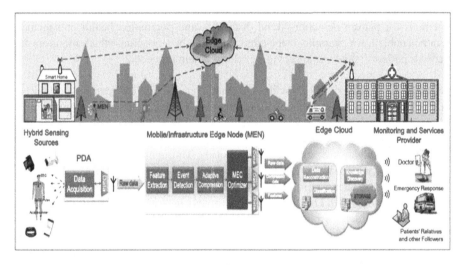

FIGURE 14.3 Edge computing framework for smart healthcare

IP cameras, smartphones, and external medical devices. All such devices are leveraged for monitoring patients' state within the smart assisted environment, which facilitates continuous-remote monitoring and automatic detection of emergency conditions. These hybrid sources of information are attached to a mobile/infrastructure edge node to be locally processed and analyzed before sending it to the cloud [17,19].

Patient Data Aggregator (PDA): Typically, the wireless Body Area Network (BAN) consists of several sensor nodes that measure different vital signs, and a PDA that aggregates the data collected by a BAN and transmits it to the network infrastructure.

Thus, the PDA is working as a communication hub that is deployed near to the patient to transfer the gathered medical data to the infrastructure [17].

Mobile/Infrastructure Edge Node (MEN): Herein, a MEN implements intermediate processing and storage functions between the data sources and the cloud. The MEN fuses the medical and non-medical data from different sources, performs in-network processing on the gathered data, classification and emergency notification, extracts information of interest, and forwards the processed data or extracted information to the cloud. Importantly, various healthcare-related applications (apps) can be implemented in the MEN, for example, for long-term chronic disease management. Such apps can help patients to actively participate in their treatment and to ubiquitously interact with their doctors anytime and anywhere. Furthermore, with a MEN running specialized context-aware processing, various data sources can be connected and managed easily near the patient, while optimizing data delivery based on the context (i.e., data type, supported application, and patient's state) and wireless network conditions [17].

Edge Cloud: It is a local edge cloud where data storage, sophisticated data analysis methods for pattern detection, trend discovery, and population health management can be enabled. An example of the edge cloud can be a hospital, which monitors and records patients' state while providing required help if needed.

Monitoring and Services Provider: A health service provider can be a doctor, an intelligent ambulance, or even a patient's relative, who provides preventive, curative, emergency, or rehabilitative healthcare services to the patients.

Edge computing contributes to improving healthcare standards by providing faster and more comprehensive treatment ubiquitously. Through large-scale deployment of health sensors, patient visits to hospitals and clinics can be reduced, especially through deploying devices that can provide computing capabilities for diagnosis of disease and patient monitoring [60]. These edge sensor devices can be easily maintained by patients and lead to new data insights on health care through their continuous monitoring of vital signs. Computing on the edge can also lower data transmission costs by migrating necessary data from the servers to the edge but edge computing faces resource allocation problems in different layers, such as CPU cycle frequency, access jurisdiction, radiofrequency, bandwidth, and so on. As a result, it has great demands on various powerful optimization tools to enhance system efficiency [11,12].

AI technologies are capable of handling this task. Essentially, AI models extract unconstrained optimization problems from real scenarios and then find the asymptotically optimal solutions iteratively with Stochastic Gradient Descent (SGD) methods. Either statistical learning methods or deep learning methods can offer help and advice for the edge [4].

14.4 ARTIFICIAL INTELLIGENCE FOR SMART HEALTHCARE

Artificial intelligence (AI) aims to mimic human cognitive functions. The rapidly growing accessibility of healthcare medical data and also the advances of big data diagnostic techniques has completed the potential of the current successful uses of AI in the healthcare system. AI is being used or trialed for a variety of healthcare and research purposes, including health and wellness, cancer treatment, dentistry, medical diagnosis, radiology tools, pathology, smart devices, and surgery. An AI system can extract useful information from a large patient population to assist with making real-time inferences for health risk alerts and health outcome prediction.

Artificial intelligence-based multiple convolutional neural network technology can simulate a doctor's cognitive, thinking, reasoning, and learning process, and promote medical advisor systems with certain intelligent auxiliary diagnostic capabilities, which can effectively make up for the lack of medical human resources, reduce costs and improve accuracy [15].

Popular AI techniques include machine learning methods for structured data, such as the classical support vector machine and neural network, and the modern deep learning, as well as natural language processing for unstructured data [48].

14.4.1 Deep Learning Algorithm

Deep learning is a modern extension of the classical neural network technique. One can view deep learning as a neural network with many layers as shown in Figure 14.4. Deep learning (also known as deep structured learning) is part of a broader family of machine learning methods based on artificial neural networks with representation learning. It is different from traditional machine learning in how representations are learned from the raw data. In fact, deep learning allows computational models that are composed of multiple processing layers based on neural networks to learn representations of data with multiple levels of abstraction [63].

In recent years, deep learning has become an important methodology in many informatics fields such as vision recognition, natural language processing, and bioinformatics. It is also a strong analytic tool for huge volumes of data. In the IoT, one open problem is how to reliably mine real-world IoT data from a noisy and complex environment that confuses conventional machine learning techniques.

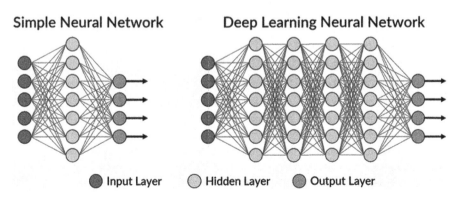

FIGURE 14.4 Deep learning neural network

So, deep learning is considered as the most promising approach to solving this problem [49].

The deep learning method includes both supervised and unsupervised learning paradigms from the layers of Artificial Neural Network (ANN). Although a number of ML algorithms exist, the recent researchers are showing extreme interest in the DL algorithm because of its self-feature extraction capability. Since the ANN adopted by deep learning model typically consists of a series of layers, the model is called a deep neural network (DNN) [2]. The taxonomy of DNN primarily includes multi-layer perceptron's (MLP), CNN, recurrent neural networks (RNN).They enable learning of task-adapted feature representations from the data [3,50].

Convolutional neural networks (CNN) are specialized neural networks for processing data that has grid-like topology (time-series data or images). The typical layer of CNN consists of three stages. The first stage performs several convolutions in parallel to produce linear activation. In the second, detector stage, each linear activation is run through nonlinear activation function. The last, third stage uses pooling function to make the output invariant to the noise and disorder. The best performances of CNN are achieved for two-dimensional image topology, while for processing of one-dimensional time series data, it is best to use RNN [3,7].

While CNNs are designed to deal with data arranged in multiple arrays, RNNs were proposed to deal with time series. RNNs make use of a special neuron with recurrent connection. This connection acts as a memory, allowing the RNN to learn the temporal dynamicity of sequential data. Each output of RNN depends on the previous outputs. RNN shares their weight factors through a very deep computation graph [7][38].

Long Short Term Memory (LSTM) is a type of RNN that uses special units to include a "memory cell" that can maintain information in memory for long periods of time. LSTM, a variant of RNN, uses a memory block inspired by a computer memory cell, where context-dependent input, output, and forget gates control what is written, read and kept in the cell in each time-step. Hence, it becomes convenient for the network to store a given input over many time steps, in effect helping LSTM layers to capture temporal properties [14,50].

The combination of smart AI software paired with reliable, more powerful computer hardware allows for medical technology to incorporate deep learning, to ultimately lead to better patient care and medical outcomes of the patients.

14.4.2 Use Cases Scenario in Healthcare

14.4.2.1 Human Activity Recognition

Human activity recognition, or HAR for short, is a broad field of study concerned with identifying the specific movement or action of a person based on sensor data. It can recognize human activities or gestures through computer system. Identified signals can be obtained from different types of detectors, such as audio sensors, image sensors, barometers, and accelerometers [12]. Movements are often typical activities performed indoors, such as walking, talking, standing, and sitting. Human Activity Recognition (HAR), e.g., generally exploits time-series data from inertial sensors to identify the actions being performed by human [9,32].

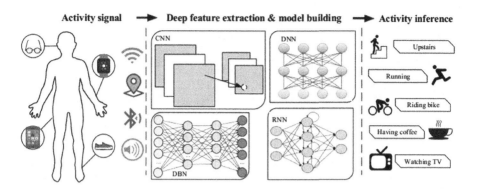

FIGURE 14.5   Human activity recognition using deep learning

In healthcare, inertial sensor data can be used for monitoring the onset of diseases as well as the efficacy of treatment options. For patients with neurodegenerative diseases, such as Parkinson's, Hunan Activity Recognition can be used to compile diaries of their daily activities and detect episodes such as freezing-of-gait events, for assessing the patient's condition [29,31,34].

To provide an accurate estimate of a wide range of activities from smartwatches, active deep learning algorithms can be utilized to provide improvements to the activity recognition model [29]. Models such as convolutional neural networks (CNN), and recurrent neural networks (RNN) employ a data-driven approach to learn discriminating features from raw sensor data to infer complex, sequential, and contextual information in a hierarchical manner as shown in Figure 14.5. They are highly suited for exploiting temporal correlations in datasets, which makes them suitable for applications such as human activity recognition (HAR) classification, where potentially a large amount of data is available; human movements are encoded in a sequence of successive samples in time; and the current activity is not defined by one small window of data alone [44].

14.4.2.2   Fall Detection

As the global aging population continues to increase, more and more attention has been paid to the damage of falls on the health of elderly. According to the data from the World Health Organization, about 28~35% of people over the age of 65 fall every year. For people over the age of 70, this number has increased to around 32~42%. Therefore, an efficient and accurate algorithm of fall detection is of great social significance [26,38].

Fall detection techniques require a continuous sensor monitoring process (several times per second) that may demand a high power consumption if the data is processed externally (in order to obtain better results); but, if the detection is done inside the embedded system itself (to reduce power consumption), the detection algorithm may reduce the fall detection accuracy and the system could have high response times if the algorithm implemented is computationally expensive [11, 38].

(a)   (b)   (c)

FIGURE 14.6   Real-time automated fall detection

ML techniques have been used to detect the fall, such as SVM, k-nearest neighbors (K- NN), random forest, or decision trees (DT) in the majority of these systems. However, these systems do not provide an acceptable precision and accuracy. Consequently, a deep learning approach is considered to improve accuracy and precision. Recurrent neural networks (RNN) are commonly employed for time sequences inputs. RNN uses a previous time input as well as the current input to make predictions. To do so, the RNN uses memories to preserve sequential information in a hidden state. Currently, there are two relevant RNN variants, long short-term memory units (LSTM) and gated recurrent unit (GRU) [39]. A fall system based on the IoT edge gateway is implemented and tested in a home as shown in Figure 14.6.

14.5   EDGE ARTIFICIAL INTELLIGENCE—DEEP LEARNING ON EDGE

Considering that AI is functionally necessary for the quick analysis of huge volumes of data and extracting insights, there exists a strong demand to integrate edge computing and AI, which gives rise to edge intelligence. Edge intelligence is not the simple combination of edge computing and AI. The subject of edge intelligence is tremendous and enormously sophisticated, covering many concepts and technologies, which are interwoven in a complex manner [4].

Edge AI means that AI algorithms are processed locally on a hardware device. The algorithms are using data (sensor data or signals) that are created on the device. A device using edge AI does not need to be connected in order to work properly, it can process data and take decisions independently without a connection. In order to use edge AI, you need a device comprising a microprocessor and sensors [65]. In edge AI, the inference algorithm is run close to the sensor that generates the data. This is different from typical AI where the sensor needs to send the data to a server for analysis [1].

The principal idea of edge computing is to sink the cloud computing capability from the network core to the network edges (e.g., base stations and WLAN) in close proximity to end devices. This novel feature enables computation-intensive and latency-critical DNN-based applications to be executed in a real-time responsive manner (i.e., edge intelligence).By leveraging edge computing, we can design an on-demand low-latency DNNs inference framework for supporting real-time edge AI applications [2].

Edge computing provides AI with a heterogeneous platform full of rich capabilities. Now a days, It is gradually becoming possible that AI chips with computational acceleration such as field programmable gate arrays (FPGAs), graphics processing units (GPUs), tensor processing units (TPUs) and neural processing units (NPUs) are integrated with intelligent mobile devices. These edge AI chips will likely find their way into an increasing number of consumer devices, such as high-end smartphones, tablets, smart speakers and wearables [1].

Sensors and devices enable continuous patient monitoring—alerting healthcare providers to clinically meaningful changes and predicting potential candidates for early intervention. Examples include smart mirrors that use intelligent hardware and software to identify subtle yet clinically relevant changes in physique and appearance, FDA-approved devices and sensors that provide in-home chronic disease monitoring, and consumer wearables such as smartwatches that track wellness indicators [66].

For a real scenario, an elderly person wearing a watch that can detect falls is a solution based on edge AI. The edge AI system use accelerometer data in real time as input to the AI algorithm that will detect when the person is falling. The watch will only connect to the cloud when it has detected a fall. One of the key properties in the example above is to have a long battery life. If the system relied on processing in the cloud it would need Bluetooth connection enabled all the time and the battery would be drained in no time [26].

While edge computing can provide latency, scalability, and privacy, there are several major challenges for deep learning at the edge. One major challenge is accommodating the high resource requirements of deep learning on less powerful edge compute resources [37]. Deep learning needs to execute on a variety of edge devices, ranging from reasonably provisioned edge servers equipped with a GPU, to smartphones with mobile processors, to barebones Raspberry as shown in Figure 14.7.

Although the edge server can accelerate DNN processing, it is not always necessary to have the edge devices executing DNNs on the edge servers instead of it intelligent

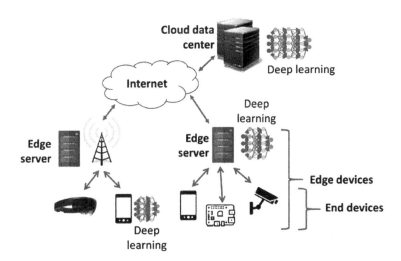

FIGURE 14.7   Deep learning on edge devices

offloading can be used. There are four different offloading scenarios: (1) binary offloading of DNN computation, where the decision is whether to offload the entire DNN or not; (2) partial offloading of partitioned DNNs, where the decision is what fraction of the DNN computations should be offloaded; (3) hierarchical architectures where offloading is performed across a combination of edge devices, edge servers, and cloud; and (4) distributed computing approaches where the DNN computation is distributed across multiple peer devices [37].

The key issue of extending DL from the cloud to the edge of the network is: under the multiple constraints of networking, communication, computing power, and energy consumption, how to devise and develop edge computing architecture to achieve the best performance of DL training and inference. As the computing power of the edge increases, edge intelligence will become common, and intelligent edge will play an important supporting role to improve the performance of edge intelligence [1].

14.5.1 Edge Intelligence Framework for Smart Healthcare

For edge AI, the main focus is on how to realize edge intelligence in a systematic way. There exist three key components in AI, i.e. data, model/algorithm and computation. A complete process of implementing AI applications involves data collection and management, model training, and model inference. Computation plays an essential role throughout the whole process [5].

There is a need to deal with edge computing architecture so that processing can be done at the devices (i.e., end-nodes), or at the gateways (i.e., edge nodes). This will reduce unnecessary data traffic and processing latency. In a healthcare scenario, end-nodes can be distinguished in a further taxonomy [16].

i. **Medical devices**: Any device intended to be used for medical purposes, such as the diagnosis, prevention, monitoring, treatment, alleviation, or compensation of a disease or an injury.

ii. **Ambient devices**: Any type of consumer electronics, characterized by their ability to be perceived at-a-glance, such as motion sensors, cameras, smoke sensors, smart appliances, etc.

iii. **Interactive devices**: Any mobile or stationary hardware component that enables the interaction between the human user and an application or the environment of the user, such as smartphones, speech recognition devices, wearable devices, etc.

In order to successfully deploy intelligent edge, the following components need to exist in the device:

i. **Connectivity**: The devices must be able to connect to networks that enable data exchange such as the internet or an internal decentralized network.

ii. **Computing**: The devices must be equipped with internal computing resources such as processing chips, enabling processing and analyzing data in near real time.

iii. **Controllability**: The devices must be capable of using databases to implement decisions such as controlling devices, making instantaneous changes, and instigating actions across networks.

iv. **Autonomy**: The devices must have autonomous computing processes and capabilities that are enabled by edge databases and that won't require assistance for monitoring, managing, and transferring data.

These require developments in the following key areas of:

1. **Hardware with partitioned edge compute**: Hardware with partitioned edge computing is providing the desired levels of connectivity, sensing, data aggregation—making up the IoT stack.

2. **Seamless cloud connectivity**: Systems with edge intelligence will need to be connected to the cloud to leverage storage and compute resources on edge.

3. **Remote device management**: Remote device management will be needed for security, management automation, edge intelligence, and API Integration.

In this Figure 14.8, a hybrid architecture is developed that will collect data from IoT devices on the edge. For that purpose, an edge computing layer is deployed between cloud platforms and health monitoring devices. The health-monitoring device transmits the monitored data to an edge device in real time through a local area network (e.g. WLAN, Bluetooth, ZigBee). And the edge device performs real-time diagnosis after receiving the data. In this framework, deep learning technology is deployed on smart devices that act as edge devices and provide real-time diagnostic capabilities. The edge layer coordinates and monitors the patient's care at home through active interaction designed to detect the care needs and critical factors involved in the care [15].

In addition to that, this framework provides the operators involved with the information about the patient by opening a clinical record (medical history, clinical diary, treatment, vital parameters, etc.). The cloud layer acts as an intermediary: it receives any requests and/or an alert sent by the edge layer after the measurement of specific vital parameters, and activates specific operating protocols [15]. This architecture works properly on resource-constrained smart devices and is deployed on smartphones that act as edge devices. It can significantly reduce diagnosis latency and network I/O, ease the pressure on the cloud platform for large user groups and large-scale data, and drastically reduce the cost for operators to build and maintain cloud platforms.

The benefits of edge intelligence are obvious—it paves the way for the last mile of AI and to provide high-efficient intelligent services for people that significantly lessens the dependency on central cloud servers and can effectively protect data privacy. However, there are still some unsolved open challenges in realizing edge intelligence challenges that include data scarcity at edge, data consistency on edge devices, bad adaptability of statically trained model, privacy and security issues, and Incentive mechanism [5].

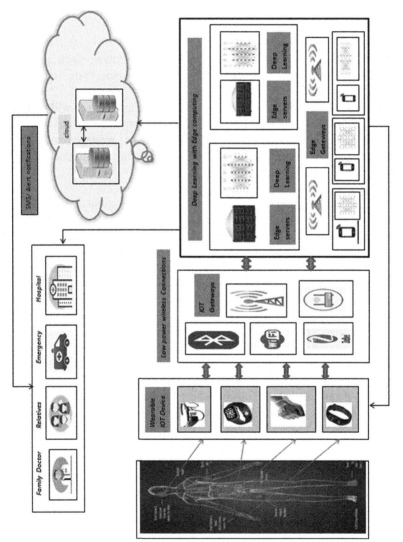

FIGURE 14.8 Edge artificial intelligence deployed with wearable IoT devices

14.6 CONCLUSION

The healthcare landscape has changed and is still changing. Patients are starting to embrace the change and using medical IoT devices to manage their health requirements. Nowadays, healthcare providers are starting to incorporate connected healthcare to drive excellence and improve treatment outcomes to give patients better healthcare experience while medical device makers are developing solutions that are more accurate, intelligent, and personalized. Ultimately, leveraging technologies in an effort to improve treatment outcomes, management of drugs and diseases and the patient experience will definitely lead to a more efficient healthcare system. However, edge intelligence is still in its primary stage and has attracted more and more researchers and companies to get involved in studying and using it. In summary, this chapter attempts to provide possible research opportunities about AI on edge in smart healthcare. Finally, some challenges and opportunities of deploying edge computing with deep learning in wearable devices are also highlighted.

BIBLIOGRAPHY

[1]   Wang, X., Han, Y., Leung, V. C. M., Niyato, D., Yan, X., and Chen, X. (2020). Convergence of Edge Computing and Deep Learning: A Comprehensive Survey. *IEEE Communications Surveys and Tutorials*, 1–1. https://doi.org/10.1109/comst. 2020.2970550

[2]   Li, E., Zeng, L., Zhou, Z., and Chen, X. (2020). Edge AI: On-demand accelerating deep neural network inference via edge computing. *IEEE Transactions on Wireless Communications*, 19(1), pp. 447–457. https://doi.org/10.1109/TWC.2019.2946140

[3]   Zhou, Z., Chen, X., Li, E., Zeng, L., Luo, K., and Zhang, J. (2019). Edge intelligence: Paving the last mile of artificial intelligence with edge computing. *Proceedings of the IEEE*, 107(8). pp. 1738–1762. https://doi.org/10.1109/JPROC.2019.2918951

[4]   Deng, S., Zhao, H., Fang, W., Yin, J., Dustdar, S., and Zomaya, A. (2019). Edge intelligence: The confluence of edge computing and artificial intelligence. *IEEE Internet of Things*, 7(8), pp. 7457–7469. doi: 10.1109/JIOT.2020.2984887.

[5]   Xu. D., Li, T., Li, Y., Su, X., Tarkoma, S., Jiang, T., Crowcroft, J., . and Hui, P. (2020) Edge intelligence: Architectures, challenges, and applications. *Proceedings of the IEEE*, arXiv:2003.12172v2.

[6]   Guk, K., Han, Sang, Lim, H., Jeong, J., Yang, J., and Jung, S.-C. (2019). Evolution of wearable devices with real-time disease monitoring for personalized healthcare. Nanomaterials *(Basel)*, 9(6), p. 813. 10.3390/nano9060813.

[7]   Shrestha, A., and Mahmood, A. (2019). Review of deep learning algorithms and architectures. IEEE Access, pp. 1–1. 10.1109/ACCESS.2019.2912200.

[8]   Khan, W. Z., Ahmed, E., Hakak, S., Yaqoob, I., and Ahmed, A. (2019). Edge computing: A survey. *Future Generation Computer Systems*, 97, pp. 219–235. https://doi. org/10.1016/j.future.2019.02.050

[9]   Uddin, M. Z. (2019). A wearable sensor-based activity prediction system to facilitate edge computing in smart healthcare system. *Journal of Parallel and Distributed Computing*, 123, pp. 46–53. https://doi.org/10.1016/j.jpdc.2018.08.010

[10] Din, S., and Paul, A. (2018). Smart health monitoring and management system: Toward autonomous wearable sensing for Internet of Things using big data analytics. *Future Generation Computer Systems*, 91, pp. 611–619. 10.1016/j.future.2017.12.059.

[11] Sarabia-Jacome, D., Lacalle, I., Palau, C. E., and Esteve, M. (2019). Efficient deployment of predictive analytics in edge gateways: Fall detection scenario. *IEEE 5th World Forum on Internet of Things, WF-IoT 2019—Conference Proceedings*, pp. 41–46. https://doi.org/10.1109/WF-IoT.2019.8767231

[12] Cooney, N., Joshi, K., and Minhas, A. (2018). A wearable Internet of Things based system with edge computing for real-time human activity tracking. *5th Asia-Pacific World Congress on Computer Science and Engineering (APWC on CSE)*, pp. 26–31. 10.1109/APWConCSE.2018.00013.

[13] Dias, D., and Paulo Silva Cunha, J. (2018). Wearable health devices—Vital sign monitoring, systems and technologies. *Sensors,* 18, p. 2414.

[14] Zebin, T., Balaban, E., Ozanyan, K. B., Casson, A. J., and Peek, N. (2019). Implementation of a batch normalized deep LSTM recurrent network on a smartphone for human activity recognition. In *2019 IEEE EMBS International Conference on Biomedical and Health Informatics, BHI 2019 – Proceedings*, pp. 1–4. https://doi.org/10.1109/BHI.2019.8834480

[15] Yu, J., Fu, B., Cao, A., He, Z., and Wu, D. (2018). EdgeCNN: A hybrid architecture for agile learning of healthcare data from IoT devices. *IEEE 24th International Conference on Parallel and Distributed Systems (ICPADS)*, pp. 852–859. 10.1109/PADSW.2018.8644604.

[16] Pazienza, A., Mallardi, G., Fasciano, C., and Vitulano, F. (2019). Artificial intelligence on edge computing: Healthcare scenario in ambient assisted living. *18 International Conference of the Italian Association for artificial Intelligence,* pp. 22–37.

[17] Abdellatif, A., Mohamed, A., Chiasserini, C.-F., Tlili, M., and Erbad, A. (2019). Edge computing for smart health: Context-aware approaches, opportunities, and challenges. *IEEE Network.* 10.1109/MNET.2019.1800083.

[18] Baker, S., Xiang, W., and Atkinson, I. (2017). Internet of Things for smart healthcare: Technologies, challenges, and opportunities. IEEE Access. 10.1109/ACCESS.2017.2775180.

[19] Pasquale, P., Aloi, G., Gravina, R., Caliciuri, G., Fortino, G., and Liotta, A. (2018). An edge-based architecture to support efficient applications for healthcare industry 4.0. *IEEE Transactions on Industrial Informatics.* 10.1109/TII.2018.2843169.

[20] Bujari, A., Gaggi, O., and Quadrio, G. (2018). Smart wearable sensors: Analysis of a real case study. *IEEE 29th Annual International Symposium on Personal, Indoor and Mobile Radio Communications (PIMRC)*, pp. 37–41. 10.1109/PIMRC.2018.8580729.

[21] Ai, Y., Peng, M., and Zhang, K. (2017). Edge cloud computing technologies for internet of things: A primer. Digital Communications and Networks, 4. 10.1016/j.dcan.2017.07.001.

[22] Lomotey, R., Pry, J., Sriramoju, S., Kaku, E., and Deters, R. (2017). Wearable IoT data architecture. *IEEE World Congress on Services (SERVICES)*, pp. 44–50. 10.1109/SERVICES.2017.17.

[23] Baker, S. B., Xiang, W., and Atkinson, I. (2017). Internet of Things for smart healthcare: Technologies, challenges, and opportunities. IEEE Access, 5, pp. 26521–26544. https://doi.org/10.1109/ACCESS.2017.2775180

[24] Manogaran, G., Shakeel, P. M., Fouad, H., Nam, Y., Baskar, S., Chilamkurti, G., and Sundarasekar, R. (2019). Wearable IoT smart-log patch: An edge computing-based Bayesian deep learning network system for multi access physical monitoring system. *Sensors*, 19, p. 3030. 10.3390/s19133030.

[25] Alam, M. M., Malik, H., Khan, M. I., Pardy, T., Kuusik, A., and Le Moullec, Y. (2018). A survey on the roles of communication technologies in IoT-based personalized healthcare applications. IEEE Access, 6, pp. 36611–36631. 10.1109/ACCESS.2018.2853148.

[26] Peña Queralta, J., Nguyen gia, T., Tenhunen, H., and Westerlund, T. (2019). Edge-AI in LoRa-based health monitoring: Fall detection system with fog computing and LSTM recurrent neural networks. *42nd International Conference on Telecommunications and Signal Processing (TSP)*, pp. 601–604, 10.1109/TSP.2019.8768883.

[27] Salkic, S., Ustundag, B. C., Uzunovic, T., and Golubovic, E. (2020). *Edge Computing Framework for Wearable Sensor-Based Human Activity Recognition. Advanced Technologies, Systems, and Applications IV—Proceedings of the International Symposium on Innovative and Interdisciplinary Applications of Advanced Technologies*,83https://doi.org/10.1007/978-3-030-24986-1

[28] Hartmann, M., Hashmi, U., and Imran, A. (2019). Edge computing in smart health care systems: Review, challenges, and research directions. *Transactions on Emerging Telecommunications Technologies*. 10.1002/ett.3710.

[29] Takiddeen, N., and Zualkernan, I. (2019). Smartwatches as IoT edge devices: A framework and survey. *Fourth International Conference on Fog and Mobile Edge Computing (FMEC)*, pp. 216–222. 10.1109/FMEC.2019.8795338.

[30] Madukwe, K., Ezika-nee Anarado, I., and Iloanusi, O. (2017) Leveraging edge analysis for Internet of Things based healthcare solutions. 2017 *IEEE 3rd International Conference on Electro-Technology for National Development (NIGERCON)*, pp. 720–725. 10.1109/NIGERCON.2017.8281940.

[31] Kwon, M. C., Ju, M., and Choi, S. (2017). Classification of various daily behaviors using deep learning and smart watch. *Ninth International Conference on Ubiquitous and Future Networks, ICUFN*, Milan, pp. 735–740. https://doi.org/10.1109/ICUFN.2017.7993888

[32] Fekri, M., and Shafiq, M. O. (2018). Deep convolutional neural network learning for activity recognition using real-life sensor's data in smart devices. *2018 IEEE 20th International Conference on E-Health Networking, Applications and Services, Healthcom 2018*, Ostrava, 1–6. https://doi.org/10.1109/HealthCom.2018.8531150

[33] Naeini, E. K., Shahhosseini, S., Subramanian, A., Yin, T., Rahmani, A. M., and Dutt, N. D. (2019). An edge-assisted and smart system for real-time pain monitoring. *2019 IEEE/ACM International Conference on Connected Health: Applications, Systems and Engineering Technologies (CHASE)*, Arlington, VA, USA, pp. 47–52.

[34] Ravì, D., Wong, C., Lo, B., and Yang, G.-Z. (2016). A deep learning approach to on-node sensor data analytics for mobile or wearable devices. 21(1), 10.1109/JBHI.2016.2633287.

[35] Pace, P., Aloi, G., Gravina, R., Caliciuri, G., Fortino, G., and Liotta, A. (2019). An edge-based architecture to support efficient applications for healthcare industry 4.0. *IEEE Transactions on Industrial Informatics*, 15(1), pp. 481–489. https://doi.org/10.1109/TII.2018.2843169

[36] Miotto, R., Wang, F., Wang, S., and Jiang, X. (2017). Deep learning for healthcare: review, opportunities and challenges. *Briefings in Bioinformatics*, 19. 10.1093/bib/bbx044.

[37] Chen, J., and Ran, X. (2019). Deep learning with edge computing: A review. *Proceedings of the IEEE*,107(8), pp. 1655–1674. 10.1109/JPROC.2019.2921977.

[38] Luna-Perejón F., Domínguez-Morales, M.J., and Civit-Balcells, A. (2019). Wearable fall detector using recurrent neural networks. *Sensors*, 19(22), p. 4885. doi:10.3390/s19224885

[39] Janković, M., Savić, A., Novičić, M., and Popović, M. (2018). Deep learning approaches for human activity recognition using wearable technology. *Medicinski Podmladak*, 69(3), 14–24. 10.5937/mp69-18039.

[40] Majumder, S., and Deen, M. J. (2019). Smartphone sensors for health monitoring and diagnosis. *Sensors,* 19(9), p. 2164. https://doi.org/10.3390/s19092164

[41] Peddoju, S. K., Upadhyay, H., and Bhansali, S. (2019). Health monitoring with low power IoT devices using anomaly detection algorithm. *2019 4th International Conference on Fog and Mobile Edge Computing, FMEC 2019*, Rome, Italy, pp. 278–282. https://doi.org/10.1109/FMEC.2019.8795327

[42] Zebin, T., Balaban, E., Ozanyan, K. B., Casson, A. J., and Peek, N. (2019). Implementation of a batch normalized deep LSTM recurrent network on a smartphone for human activity recognition. *2019 IEEE EMBS International Conference on Biomedical and Health Informatics, BHI 2019 – Proceedings*, Chicago, IL, USA, pp. 1–4. https://doi.org/10.1109/BHI.2019.8834480.

[43] Rodrigues, J., Segundo, D., Arantes Junqueira, H., Sabino, M., Prince, R., Al-Muhtadi, J., and Albuquerque, V. (2018). Enabling technologies for the internet of health things. *IEEE Access*. 10.1109/ACCESS.2017.2789329.

[44] Zebin, T., Scully, P., Peek, N., Casson, A., and Ozanyan, K. (2019). Design and implementation of a convolutional neural network on an edge computing smartphone for human activity recognition. IEEE Access. 10.1109/ACCESS.2019.2941836.

[45] Hiremath, S., Yang, G., and Mankodiya, K. (2014). Wearable Internet of Things: Concept architectural components and promises for person-centered healthcare. *2014 4th International Conference on Wireless Mobile Communication and Healthcare—Transforming Healthcare Through Innovations in Mobile and Wireless Technologies (MOBIHEALTH)*, Athens, Pp. 304–307. 10.4108/icst.mobihealth.2014.257440.

[46] Li, J., Ma, Q., Chan, A., and Man, S. S. (2019). Health monitoring through wearable technologies for older adults: Smart wearables acceptance model. *Applied Ergonomics*, 75, pp 162–169. 10.1016/j.apergo.2018.10.006.

[47] Haghi, M., Thurow, K., and Stoll, R. (2017). Wearable devices in medical Internet of Things: Scientific research and commercially available devices. *Healthcare Informatics Research*, 23(1), pp. 4–15. https://doi.org/10.4258/hir.2017.23.1.4.

[48] Jiang, F, Jiang, Y., Zhi, H., et al. (2017). Artificial intelligence in healthcare: Past, present and future. *Stroke Vascular Neurology,* 2(4), pp. 230–243. doi:10.1136/svn-2017-000101.

[49] Li, H., Ota, K., and Dong, M. (2018). Learning IoT in edge: Deep learning for the Internet of Things with edge computing. *IEEE Network*, 32, pp. 96–101. 10.1109/MNET.2018.1700202.

[50] Biswas, D., Everson, L., Liu, M., Panwar, M., Verhoef, B. E., Patki, S., and Van Helleputte, N. (2019). CorNET: Deep learning framework for PPG-based heart rate estimation and biometric identification in ambulant environment. *IEEE Transactions on Biomedical Circuits and Systems*, 13(2), pp. 282–291. https://doi.org/10.1109/TBCAS.2019.2892297

[51] Wearable Internet of Things for healthcare. https://blog.accemy.com/2018/07/wearable-internet-of-things-for.html

[52] https://www.seeedstudio.com/blog/2020/01/20/what-is-edge-ai-and-what-is-it-used-for/

[53] Seth, B., Dalal, S., and Kumar, R. (2019) Securing bioinformatics cloud for big data: Budding buzzword or a glance of the future. In: Kumar, R., and Wiil, U. (eds) *Recent Advances in Computational Intelligence,* Studies in Computational Intelligence, vol. 823. Cham: Springer. https://doi.org/10.1007/978-3-030-12500-4_8

[54] www.prnewswire.com/in/news-releases/iot-healthcare-market-worth-15807-billion-usd-by-2022-619971033.html

[55] www.starhub.com/business/resources/blog/healthcare-is-rapidly-transforming-with-ai-and-iot.html

[56] https://vilmate.com/blog/why-use-ai-enabled-iot-in-healthcare/

[57] Seth, B., and Dalal, S. (2018) Analytical assessment of security mechanisms of cloud environment. In Saeed, K., Chaki, N., Pati, B., Bakshi, S., and Mohapatra, D. (eds),

Progress in Advanced Computing and Intelligent Engineering, Advances in Intelligent Systems and Computing, vol, 563. Singapore: Springer. https://doi.org/10.1007/978-981-10-6872-0_20

[58] https://www.networkworld.com/article/3224893/what-is-edge-computing-and-how-it-s-changing-the-network.html

[59] Seth, B., Dalal, S., and Kumar, R. (2019) Hybrid homomorphic encryption scheme for secure cloud data storage. In Kumar, R., and Wiil U. (eds) *Recent Advances in Computational Intelligence*, Studies in Computational Intelligence, vol. 823. Cham: Springer. https://doi.org/10.1007/978-3-030-12500-4_5

[60] Edge computing and healthcare: Looking to the future. www.healthitoutcomes.com/doc/edge-computing-and-healthcare-looking-to-the-future-0001

[61] Will edge computing transform healthcare. https://healthtechmagazine.net/article/2019/08/will-edge-computing-transform-healthcare

[62] Electronic T-shirt: How to make a sensor-filled shirt https://maker.pro/pcb/projects/e-tshirt-diy-tutorial.

[63] Hexoskin smart garments specifications. www.hexoskin.com/

[64] Deep learning. https://en.wikipedia.org/wiki/Deep_learning

[65] Wearable technology: This "smart shirt" can accurately monitor lung disease https://gulfnews.com/technology/wearable-technology-this-smart-shirt-can-accurately-monitor-lung-disease-1.1569825487265

[66] What is edge AI? https://www.imagimob.com/blog/what-is-edge-ai

[67] An Rx for healthcare providers: Edge computing and IoT solutions www.cio.com/article/3526606/an-rx-for-healthcare-providers-edge-computing-and-iot-solutions.html

[68] Xu. D., Li, T., Li, Y., Su, X., Tarkoma, S., Jiang, T., Crowcroft, J., and Hui, P. (2020) Edge intelligence: Architectures, challenges, and applications. *Proceedings of the IEEE*, arXiv:2003.12172v2

[69] Salkic, S., Ustundag, B. C., Uzunovic, T., and Golubovic, E. (2020). Edge computing framework for wearable sensor-based human activity recognition. *Advanced Technologies, Systems, and Applications IV—Proceedings of the International Symposium on Innovative and Interdisciplinary Applications of Advanced Technologies*, 83. https://doi.org/10.1007/978-3-030-24986-1

[70] Shrestha, A., and Mahmood, A., (2019). Review of deep learning algorithms and architectures. *IEEE Access*. 10.1109/ACCESS.2019.2912200

15 An Approach to IoT Framework and Implementation of DoS Attack

*Umang Garg, Shalini Singh,
Shivashish Dhondiyal, and Neha Gupta*

CONTENTS

The Internet of Things (IoT) revolution comprised interconnected dumb devices to exchange information. The integration of modules such as RFID (Radio Frequency Identification), WSAN (Wireless Sensor and Actor-Network) and objects with the internet can provide remote access to the devices. IoT offers numerous services in various domains and utilizing the traditional internet infrastructure. IoT equipment are more vulnerable due to the involvement of the Internet and Cloud infrastructure. Currently, there is no standard framework or any well-defined structure to regularize the working of the IoT. However, several frameworks were proposed by researchers based on functionalities. In this chapter, we first discuss the most relevant protocols that can be used in the IoT domain. Second, we classify different IoT frameworks with their functionalities and limitations. Finally, a case study is performed in the IoT simulated environment and a Denial of Service attack (DoS) performed. We hope that our work will be helpful for those who have an interest in the field of IoT security.

15.1 INTRODUCTION

IoT is a key term that resides in every area of day-to-day life ranging from small wearables to large appliances used in big industrial systems [1]. IoT has the facility to communicate between human to human, device to device, and human to device. A device-to-device connection provides a special feature that can create a powerful application that is going to be utilized in numerous fields. IoT is one of the major research fields for the researchers that provides machine-to-machine interaction with the help of sensors and actuators [2]. The size of IoT is expected to reach 20 billion devices and an investment of US$1.7 trillion up to 2020 [3]. Kevin Ashton, executive director of the Auto-ID Center, initiated the term IoT in the 1990s [4]. He believed Radio Frequency Identification (RFID) is a basic requirement for the implementation of IoT.

Several applications work with the concept of IoT in our daily life. According to NIC [5], by 2025 internet nodes will also be residing in almost all things like furniture, documents, etc. There are several advantages of IoT systems, which enhance device communication, provide excellent automation and control, embedded with more technical features, and enable powerful surveillance features to be performed more efficiently. As one of the emerging fields, IoT has faced several threats and loopholes such as authentication, trust management, various attacks and malware, etc. IoT security is one of the major issues that can attract researchers as well as industry.

Many research organizations are evolving to set up and design a secured architecture for IoT-based systems. To provide a reliable service [6], there are some architectures proposed for IoT such as three-layer, four-layer, five-layer, and seven-layer architectures. The three-layered architecture explained the basic idea of IoT (perception layer, internet layer, and application layer). Although this architecture was a finer aspect of IoT, it is not enough for research and security of the devices. In five-layer architecture, the transport and business layer were also introduced to define the concept in a better manner. Since the IoT has direct influence on the users' lives, there is a requirement of well-defined architecture that can deal with security threats and challenges. Recently, CISCO proposed an IoT model that consists of seven layers such as edge, communication, edge computing, data accumulation, data abstraction, application, and users and centers. However, security threats and issues are not well-recognized in the field of IoT. In this chapter, we provide a detailed review of the distinguished architecture, attacks, limitations, and their challenges in the framework of IoT. The main contributions of the paper are summarized as follows:

- To provide a detailed comparative study of different IoT architectures.
- To discuss the challenges and issues associated with IoT architecture.
- We also demonstrate a real-time Denial of Service attack scenario on the implemented testbed set-up.

The rest of the chapter can be defined as follows: Section 15.2 provides an in-depth discussion about applications and protocols. Section 15.3 analyses IoT distinguishing architectures. Section 15.4 points out challenges and open research issues. Section 15.5 defines the case study of real-time attack scenarios on an open-source attacking tool. Finally, the last section is the conclusion.

15.2 RELATED WORK

There are several suggestions to try to identify an IoT architectural pattern standard for a particular domain of applications. Gronbaek et al. [7] introduced the comparative goals and OSI like architectural model. Atzori et al. [8] introduce and compare different visions of IoT and illustrate which is best suited. Authors addressed various challenges, issues and focused on some security issues like network security, privacy, hardware security, etc. The enabling factor of the paper is to integrate several technologies and communication solutions. The author also identified different applications in various domains along with futuristic application domain.

In a similar fashion, Fuqaha et al. [9] summarized the most relevant information by comparing different survey papers from various perspectives in the IoT field and provided in-depth analysis of internet protocols, communication technologies, smart sensors, etc. Authors also compare distinguish protocols standards like IETF, IEEE, and EPCglobal. Although several research challenges were discussed in the present literature, authors did not deal with some novel approaches like the integration of IoT with cloud, big data analytics, and fog computing, etc. Atzori et al. [10] discussed a elementary three-layer architecture that worked with the perception layer, network layer, and application layer. The functioning at the first layer is to collect the data from the physical environment. A user can interact with the application to analyze and monitor the result possessed by the sensors in the application layer.

Wu et al. [21] provides an overview of IoT along with architectural design, which includes five-layer for IoT. A five-layered architecture contains edge node, object abstraction, service composition, service management, and application layer. Moreover, the fundamental architecture of the internet required modification to match the current IoT challenges. Sarkar et al. [11] suggested a distributed network-like architecture for IoT specifically for augmentation of IoT that addresses heterogeneity, which can provide different levels of cogitation to handle various issues like gullibility, preservation, interoperability. Authors also justified some objectives using real-life examples.

Gaur et al. [12] recommended an IoT architecture for a smart city exploiting a very large amount of data and information. Authors make use of a semantic modelling theory for providing customized services in a smart city environment mainly focused on only important areas as to cover each aspect is very difficult. However, refinement is needed to improve scalability and interoperability. Krco et al. [13] introduced a prevalent structure for IoT architecture and offering a framework and guidelines to support the development of specific IoT applications for system developers according to the European perspective. There are various dominant IoT applications in the perspective of market opportunity, in which the healthcare industry has the highest share with 41% while manufacturing is second with 33%. The rest of the paper gave an in-depth discussion on various protocols and applications of IoT. We deal with a comprehensive, detailed, and exhaustive survey of IoT architectures.

15.3 BACKGROUND PROTOCOLS AND APPLICATIONS

The IoT, like traditional networks, is broken up into different layers. The communication on each layer characterizes and standardizes the communication functions of a network. This section discusses a number of protocols with their characteristics and applications of IoT in brief. There are a diverse set of protocols (as shown in Figure 15.1) [14] used for IoT applications. Some of the oldest protocols like X10 were used in IoT but now we have used some of the latest protocols that are based on wireless technology such as Zigbee, Zwave, 6LoWPAN, BLE, IEEE 802.15.4, TLS, NFC, CoAP, DTLS, LoRaWAN. Here, we discuss some of them as follows:

ZigBee [15]: ZigBee technology aims to ensure effective and safe short-range communication in an IoT framework that constitutes mesh, star, and cluster tree topologies. The Zigbee network can target radio frequency applications that crave a very low data flow rate and long battery life. It provides a frequency of 2.4 GHz (worldwide).

Z-wave [16]: For the implementation of IoT applications, z-wave can provide a wireless connection protocol. Basically, z-wave can create a technology that can be used for communication in IoT. Also, the frequency of z-wave can be used for applications of less than 1 GHz.

6LoWPAN: IPv6 over Low Power Wireless Personal Area Network can be explained to the devices, which are compatible for IEEE 802.15.4. It makes use of star and mesh topology and provides a frequency band of 2.4 GHz.

LoRaWAN: It stands for Low Power Wide Area Network, which is a protocol for WAN existing at the MAC layer. It is intended to enable low-powered computers to interact over lengthy distance wireless links. It defines 64 kHz, 125 kHz channels [17]. Major benefits of LoRaWAN are repeater elimination, device cost reduction, increasing gadgets' lifetime, network capacity improvement and supporting a rich set of devices.

CoAP: CoAP can be defined as a constrained application protocol [18] that is a type of specific protocol for restricted internet applications. CoAP transactions provide reliable UDP messaging and provides Asynchronous message exchange, low overhead those are simple to parse. There are two levels for the CoAP design: the communication and request-respond layers. The communication layer has domination in the interaction through the User-Datagram protocol while the request-response layer sends the respective message retaining certain sequences that will be used to handle and prevent operational problems like message failure.

15.3.1 IoT APPLICATIONS

This section defines some existing applications in the field of IoT. Applications can perfectly integrate distributed systems and applications with normal web service protocols and service composition techniques. IoT applications empower device-to-device and user-to-device interactions reliably and robustly [19]. IoT applications on

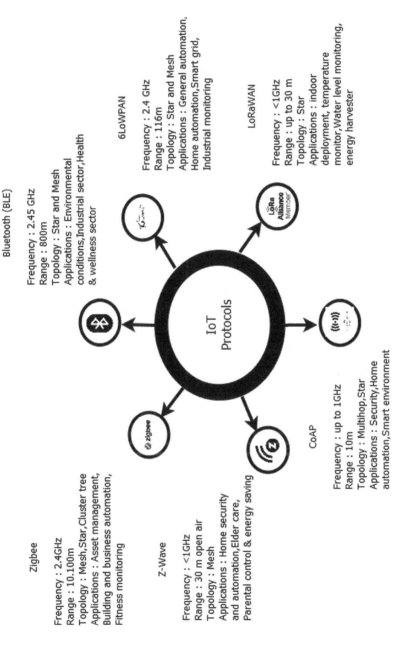

FIGURE 15.1 IoT protocols

human-to-machine confirm that data has been acquired in an actual time scenario. For example, to monitor the status of transported goods and their location, it requires data visualization to present information to humans easily and efficiently. While machine-to-machine interactions do not require data visualization, these applications should work with the intelligence so that machines can monitor, recognize and communicate with each other and resolve the problem without the need of humans. Another classification based on the technology trends and literature review can be used for business applications which include observing and supervising, enormous information and enterprise examination, and information distribution and collaboration. There are numerous steps involved in each application available for IoT. Here we classified the applications as follows:

Monitoring and control: The very first type of applications is collecting the data from sensors and analyzing performance on real-time applications. For example, Lively is an intelligent system that can be used to track and alert adults, spiroscout is a tool that fits on top of an asthma inhaler, Lapka PEM is an intelligent private environment monitor that enables the user to evaluate, retrieve and analyze the concealed characteristics of their environment.

Information sharing and collaboration: It can occur between user-to-user, user-to-device and device-to-device. Sensors are attached to devices that sense from the environment and an alert is sent to the user if some unusual activity occurs. For example, in the supermarket, sensors are put where refrigeration is required so that if any abnormal activity is discovered, the alert can be sent to the manager, Shopkick's application is used to improve data sharing that provides deals, discounts, suggestions, and benefits that remind shoppers.

Another classification of IoT applications can be gathered in the ensuing domain such as smart environment, smart healthcare, smart transportation system, smart logistics, smart utilities, etc. Applications for structural health follow-up are apps for precision farming, platforms for tracking and supervising objects.

15.4   CLASSIFICATION OF IOT ARCHITECTURES

In this section, we introduce three different types of architectures for IoT. In the first architecture, there are three layers such as application, cloud and WSN. Basically, it is an extended version of WSN, in which we can communicate with hardware and actuate them to perform any operation. Second, a five-layer architecture will facilitate support edge nodes and provide some service management to those nodes. Finally, the third model represents bidirectional data flow depends on the type of application. This model contains seven layers, in which data abstraction and data accumulation can be done mid-level of the architecture. We briefly describe the functionality of each layer of this architecture.

15.4.1   THREE-LAYER ARCHITECTURE

A basic IoT architecture can be described in hierarchical manner (as shown in Figure 15.2). It is a typical architecture consisting of three main layers. It includes the perception layer, network layer, and application layer [20].

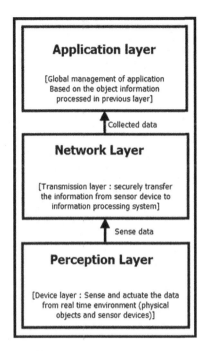

FIGURE 15.2 Three-layer architecture

Perception layer: It is the architectural foundation of three-tier structure. This layer incorporates and retains the network layer asset management. The primary role is to test environment information with different types of tools to view and differentiate the state of the systems and accountable for collecting, measuring and extracting information from physical machines. It obtains information by sensing through RFID, sensors and actuators.

Network layer: The sensitive information is fused at the network layer and transmitted via the internet, and ubiquitous networks to the respective top-layer platforms. This layer promotes computational systems like cloud services systems, gateways, portable interaction and routing systems, etc.

Application layer: This is the layer that can be placed at top of the three-layer architecture. The application layer can provide services to the end-user whenever it is required. The core issue is exchanging information with the community and ensuring data security for the application layer.

The three-layer architecture appears very simple, but the network and application layer have some complexities in pruning the information (information aggregation, information mining and analytic). Thus, there is a need for improvement in the previous architecture.

15.4.2 Five-Layer Architecture

The IoT interconnect hundreds of thousands of gadget creating more traffic and information storage. IoT will also experience many security-related difficulties. This led to the development of new architecture [10] (as shown in Figure 15.3). Here we are defining the functionality of these layers as follows:

Edge nodes: Edge nodes are the nodes at which data, signal transformation and data can be sent and received. Edge nodes may be as small as a silicon chip or as large as a car. These nodes can be used to sense and actuate the real-time environment with some physical parameters.

Object abstraction: It can transfer the information from sensors to the IoT devices safely. This level focuses on viewing and processing information to make it easier to create productive applications.

Service management: This layer can manage overall accountability of the services and establish the service that meets the enterprise requirements. Repositories of this sub-layer can be performed to determine the service object pair inside the IoT network. Service functionality may split across QoS service handling, lock accountability. It includes information processing, database linkage and ubiquitous computing.

Service composition: Service layer offers functions for the development of services provided by networked entities to build an application. A service can efficiently be handled by effective networked device relationships using work-flow languages.

Application layer: This layer can provide the interaction with the end-users. It is liable for managing the complete IoT arrangements like service, network, user confidential data and applications. Depending upon the information received form the implementation layer, application layer builds the flowcharts, designs, models and helps to determine potential customer activities.

Five-layer architecture is a conceptual process and does not work properly when nodes and traffic increase. It does not clarify and standardize the IoT framework with increasing demand. Therefore, we need a novel processing architecture that can provide correct implementation of IoT features and schemes.

15.4.3 SEVEN-LAYER ARCHITECTURE

Some existing architectures do not function properly for continuous information development, integration of manufacturing and corporate networks enabling various applications. Thus, there is a requirement to establish a new communication and processing architecture to fulfill enormous business potential. There are seven layers [22] in this IoT architecture as shown in Figure 15.4.

Physical devices and controllers: It is also known as an item in IoT/edge nodes. Multiple devices act as edge nodes that can send and receive information. Some sensors, biochip transponder, RFID, microcontroller, antennas, and many more are examples of physical devices and controllers.

Connectivity: The primary objective is the prompt transmission of data. This connectivity is done through Wi-Fi, Bluetooth, WSN, Thread, Zigbee, etc. Here the first step is authentication, a user identity verification process when connecting with a physical device, which can be done by using any of the connecting media.

Edge computing: At the edge computing layer, there is no need to transfer the information source to the cloud or other centralized processing servers. Fog computing describes how edge computing should function and enables computational activities.

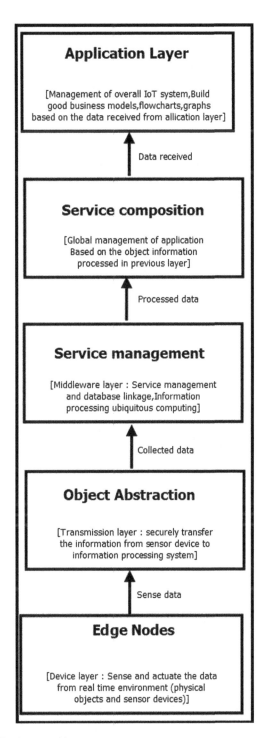

FIGURE 15.3 Five-layer architecture

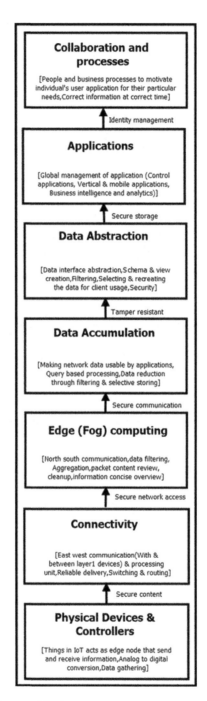

FIGURE 15.4   Seven-layer architecture

TABLE 15.1
Comparative analysis of IoT architecture

Parameter architecture	Three layer [10]	Five layer [23]	Seven layer [24]
Protocol	RFID, WISP	Bluetooth, CoAP	MQTT, Websocket
Application	Low level	Analytic, complex	Critical business
Data filtering	No	No	Yes
Node heterogeneity	No	Yes	Yes
Interface	Command	GUI	GUI
Security	No	Less	More
Content inspection	No	Yes	Yes
Latency	More	More	Minimum
Computing	Event based	Event based	Query based
Scale analytics	Terabyte	Petabyte	Exabyte
Scalability	No	Yes	Yes

The major utilization of the edge computing layer is to provide data filtering, aggregation, concise information, etc.

Data accumulation: It makes it possible to use network data for applications. It takes data and converts it into data at rest. This data can be used on a non-real time basis through applications. Whenever necessary, updated data is obtained by implementation through query-based processing by turning its format from network packets to the database relational tables.

Data abstraction: This level concentrates on displaying information and storing it to allow easier, more efficient applications to be developed. This layer is used to abstract the data for higher layers so that it can be analyzed in an optimized way.

Application: Applications must be differentiated so that processed data can properly be understood by upper layers. Applications differ according to interconnected devices, data nature, and company requirements. For example, analytical applications, system management applications, mobile applications, business applications, control center applications, and many more.

Collaboration and processes: Information generated by IoT and the IoT system itself is not as meaningful unless it produces some intervention that often involves individuals and processes. The primary objective of any application is to perform business processes to execute the user's application for their particular requirements. The application provides the correct information at the correct time so that the appropriate action can be performed.

A comparative analysis of the distinguished architectures is shown in Table 15.1 that can be categorized with different parameters.

15.5   CASE STUDY

In this section, we demonstrate the use of the open-source DoS attacking tool pentmenu that contains Recon, extraction and DoS module. The Recon module can perform a UDP scan, IPsec scan, while the DoS module can execute ICMP echo flood, TCP Syn flood, SSL DoS, and Slowloris attack. The extraction module can be used to send files using Netcat and listen to the TCP or UDP port.

15.5.1   EXPERIMENTAL TEST-BED DETAILS

To implement the DoS attacking scenario, we require a system with 8 GB main memory, and the processor belongs to the i5 family. In order to implement an IoT-based application, we use a Raspberry Pi 3b module, an ESP 8266 module for Wi-Fi connection, and some sensors to sense and actuate some data. An attacking scenario can be implemented on Kali VM.

15.5.2   ATTACK SCENARIO

Here, we perform a slowloris attack scenario that can send HTTP headers to the target host slowly using Netcat. That results in the starvation of the authorized resources. This attack is very effective if the server does not have limited time to complete HTTP requests coming from different sources.

15.5.3   EXECUTION STEPS

Step 1: First create a clone on the ID of the Git file.
Step 2: Open the terminal in the pentmenu directory and write the command ./
pentmenu (as shown in Figure 15.5).
Step 3: There are various options available for attacks such as recon, Dos, and extraction. We can select any option with target IP address, port number, and time interval (as shown in Figure 15.6).
Step 4: Now we perform the attack on IoT device and stop the services of the target server (as shown in Figure 15.7).

15.6   CONCLUSION

In this chapter, we have examined different IoT protocols and differentiate them on various parameters. This chapter concentrates on the study of different IoT applications, and table-based classification is described. We have classified three

```
root@kali:~/Desktop/pentmenu# ./pentmenu
```

FIGURE 15.5   Run the pentmenu

```
1) Recon
2) DOS
3) Extraction
4) View Readme
5) Quit
Pentmenu>2
1) TCP SYN Flood      3) TCP RST Flood      5) UDP Flood       7) Slowloris      9) Distraction Scan
2) TCP ACK Flood      4) TCP XMAS Flood     6) SSL DOS         8) IPsec DOS      10) Go back
Pentmenu>7
Using netcat for Slowloris attack....
Enter target:
172.0.0.1
Target is set to 172.0.0.1
Enter target port (defaults to 80):
80
Using Port 80
Enter number of connections to open (default 2000):
99999
Choose interval between sending headers.
Default is [r]andom, between 5 and 15 seconds, or enter interval in seconds:
5
use SSL/TLS? [y]es or [n]o (default):
y
```

FIGURE 15.6   Command with options

```
Slowloris attack ongoing ... this is connection 1058, interval is 5 seconds
Slowloris attack ongoing ... this is connection 1059, interval is 5 seconds
Slowloris attack ongoing ... this is connection 1060, interval is 5 seconds
Slowloris attack ongoing ... this is connection 1061, interval is 5 seconds
Slowloris attack ongoing ... this is connection 1062, interval is 5 seconds
Slowloris attack ongoing ... this is connection 1063, interval is 5 seconds
```

FIGURE 15.7   Attack execution

distinct IoT architectures with layers. Current architectures (three layers and five layers) do not work well with an increasing number of devices and people in terms of scalability and effectiveness. Therefore, a seven-layered architecture addresses all constraints of the prior architectures. We also implement an attacking scenario using an open-source attacking tool.

BIBLIOGRAPHY

[1]   Y. Mao, J. Zhang, C. You, K. B. Letaief, and K. Huang, "A survey on mobile edge computing: The perspective of communication," *IEEE Communications Surveys and Tutorials*, 19(4), 2017, pp. 2322–2358.

[2]   B. W. Khoueiry and M. Soleymani, "A novel machine-to-machine communication strategy using rate-less coding for the Internet of Things,", *IEEE Internet of Thing Journal*, 4662(2), 2017, pp. 937–950.

[3]   C.-W. Tsai, C.-F. Lai, L. T. Yang, and M.-C. Chiang, "Datamining for IoT: A survey," *IEEE Communications Surveys and Tutorials*, 16(1), 2013, pp. 77–97.

[4]   K. Ashton, "Internet of Things'," *RFID Journal*, 22(7), 2009, pp. 97–114.

[5]   M. binti Mohamad Noor and W. H. Hassan, "Current research on Internet of Things (IoT) security: A survey," *Comput. Networks*, vol. 148, pp. 283–294, 2019.

[6]   F. Mattern and C. Floerkemeier, "From the internet of computers to the Internet of Things," in *Active Data Management to Event-Based Systems*, Springer-Verlag Berlin Heidelber, 2010, pp. 242–259.

[7]   A. Mosenia and N. K. Jha, "A comprehensive study of security of internet-of- things," *IEEE Transactions on Emerging Topics in Computing*, 5(4), pp. 586–602, 2016.

[8]   I. Gronbaek, "Architecture for the internet of things (IoT): Api and interconnect," in *Proceedings of the 2008 Second International Conference on Sensor Technologies and Applications*, ser. SENSORCOMM '08. Washington, DC: IEEE Computer Society, 2008, pp. 802–807. Available: https://doi.org/10.1109/SENSORCOMM.2008.20.

[9]   L. Atzori, A. Iera, and G. Morabito, "The internet of things: A survey," *Computer Networks*, 10, 2010, pp. 2787–2805.

[10] A. Al-Fuqaha, M. Guizani, M. Mohammadi, M. Aledhari, and M. Ayyash, "Internet of things: A survey on enabling technologies, protocols, and applications," *IEEE Communications Surveys and Tutorials*, 17(4), 2015, pp. 2347–2376.

[11] L. Atzori, A. Iera, and G. Morabito, "The internet of things: A survey," *Computer Networks*, 54(15), 2010, pp. 2787–2805.

[12] C. Sarkar, A.U.N. S.N., R. V. Prasad, A. Rahim, R. Neisse, and G. Baldini, "Diat: A scalable distributed architecture for IoT," *IEEE Internet of Things Journal*, 2(3), 2014, pp. 230–239.

[13] A. Gaur, B. Scotney, G. Parr, and S. McClean, "Smart city architecture and its applications based on IoT," *Procedia Computer Science*, 52, 2015, pp. 1089–1094.

[14] S. Krco, B. Pokric, and F. Carrez, "Designing IoT architecture(s): A European perspective," in *2014 IEEE World Forum on Internet of Things (WF- IoT)*. IEEE, Seoul, South Korea, 2014, pp. 79–84.

[15] A. Triantafyllou, P. Sarigiannidis, and T. D. Lagkas, "Network protocols, schemes, and mechanisms for internet of things (IoT): Features, open challenges, and trends," *Wireless Communications and Mobile Computing*, vol. 2018, 2018.

[16] P. Kinney, "Zigbee technology: Wireless control that simply works," in *Communications Design Conference*, 2, 2003, pp. 1–7.

[17] M. B. Yassein, W. Mardini, and A. Khalil, "Smart homes automation using z-wave protocol," in *2016 International Conference on Engineering and MIS (ICEMIS)*, IEEE, Agadir, Morocco, 2016, pp. 1–6.

[18] B. Seth, S. Dalal, and R. Kumar, Hybrid homomorphic encryption scheme for secure cloud data storage. In Kumar, R., and Wiil, U. (eds) *Recent Advances in Computational Intelligence*, Studies in Computational Intelligence, vol. 823. Cham: Springer, 2019. https://doi.org/10.1007/978-3-030-12500-4_5

[19] M. Luvisotto, F. Tramarin, L. Vangelista, and S. Vitturi, "On the use of lo- rawan for indoor industrial IoT applications," *Wireless Communications and Mobile Computing*, 2018, pp. 1–11.

[20] A. Mali and A. Nimkar, "Security schemes for constrained application proto- col in IoT: A precise survey," in *International Symposium on Security in Computing and Communication*. Cham: Springer, 2017, pp. 134–145.

[21] I. Lee and K. Lee, "The internet of things (IoT): Applications, investments, and challenges for enterprises," *Business Horizons*, 58(4), 2015, pp. 431–440.

[22] Z. Yang, Y. Yue, Y. Yang, Y. Peng, X. Wang, and W. Liu, "Study and application on the architecture and key technologies for IoT," in *2011 International Conference on Multimedia Technology*. IEEE, Hangzhou, China , 2011, pp. 747–751.

[23] M. Wu, T.-J. Lu, F.-Y. Ling, J. Sun, and H.-Y. Du, "Research on the architecture of internet of things," in *2010 3rd International Conference on Advanced Computer Theory and Engineering (ICACTE)*, IEEE, Chengdu, China, 2010, p. V5-484.

[24] B. Seth, S. Dalal, and R. Kumar, Securing bioinformatics cloud for big data: Budding buzzword or a glance of the future. In Kumar, R., and Wiil, U. (eds) *Recent Advances in Computational Intelligence*, Studies in Computational Intelligence, vol. 823. Cham: Springer, 2019, pp. 1–27, https://doi.org/10.1007/978-3-030-12500-4_8.

[25] M. Muntjir, M. Rahul, and H. A. Alhumyani, "An analysis of internet of things (IoT): Novel architectures, modern applications, security aspects and future scope with latest case studies," *International Journal of Engineering Research Technology*, 6(6), 2017, pp. 422–447.

[26] Y. Song, S. S. Yau, R. Yu, X. Zhang, and G. Xue, "An approach to QoS-based task distribution in edge computing networks for IoT applications," in *2017 IEEE international conference on edge computing (EDGE)*. IEEE, 2017, pp. 32–39.

16 WSN Based Efficient Multi-Metric Routing for IoT Networks

Meenu Vijarania, Neeraj Dahiya, Surjeet Dalal, and Vivek Jaglan

CONTENTS

Internet of Things (IoT) is a hefty network where numerous devices are inter-connected and have the capability to communicate with each other. Nowadays, IoT is being used in many applications such as e-healthcare systems, sensing and notifying the environment, traffic management, military environment, etc. In smart environments, IoT communicates with a mobile ad-hoc network (MANET) and sensor network (WSN) making it more eye-catching to users and inexpensive. Communication between mobile ad-hoc networks and a wireless sensor with the IoT leads to the formation of a novel MANET. Such a system provides better mobility for a user and reduces the deployment expenses of the network. Energy efficiency is one of the major challenges in IoT systems as they have a limited battery lifetime. In this chapter, a novel multi-metric energy efficient algorithm has been proposed for IoT networks that selects the optimal path with minimum energy expenditure that leads to the enhanced lifetime of network and uniform energy consumption by the nodes. Simulation results show that the proposed technique outperforms the existing algorithms and extends the lifetime of the network.

16.1 INTRODUCTION

The use of the internet is increasing exponentially on a daily basis in this current era. There are an enormous number of devices that are associated with the internet and these physical devices are also referred to as the things that are using internet in a similar way as humans are connected to the internet for their requirements [1]. In the era of wireless communication, IoT is gaining huge attention. IoT comprises an array of things or objects such as mobile phones, sensors, Radio Frequency Identification (RFID) tags, actuators, etc. which have internet connectivity using both wireless and wired networks. Objects in IoT can communicate sense and transfer data with each other. They turn out to be powerful tools and are able to respond to irregularities and to emergency situation immediately. Therefore, getting insights about real-world physical processes, IoT is the eventual solution. Thus, IoT brings several new challenges problems regarding the networking aspects. In fact, IoT objects are characterized by low resources in terms of both energy capacity and computation power. Thereby, efficient resource utilization is the main concern of IoT besides the obvious scalability problems.

Many protocols have been developed to deal with the routing problem in IoT. In the literature, numerous routing protocols for IoT implement proactive, reactive and hierarchical schemes to route the data packets efficiently among IoT devices. Our idea in this chapter is to suggest a proficient routing technique that reduces the total energy consumed, enhancing network lifetime and uniform traffic distribution among IOT devices. All nodes in IoT are powered by battery. After deployment, IoT devices are kept with no human intervention for a long time. Lack of energy-efficient techniques make the entire energy deplete in a few days, which leads to the premature death of some devices and causes the network to partition.

During the communication, energy is wasted due to some other reasons like control overhead, interference and collision. Hence, various routing mechanism have been developed focusing towards IoT lifetime enhancement and minimizing power consumption.

The conventional routing mechanism such as DSR, AODV and OLSR are used to find the shortest path between any two devices. These techniques do not include load balancing and efficient energy utilization parameters. Various energy efficient algorithms have been developed, but these techniques have not taken into consideration energy efficient IoT. An example of device-to-device communication of IoT has been shown in Figure 16.1.

Wireless sensors are an important element of IoT networks, which helps to connect with remote users. These sensors gather the valuable information, convert it into digital format and send it to other devices in the network. With the growing demand and deployment of IoT sensors, the main challenges are the selection of transmission medium and packets routing in the heterogeneous networks. The correlation of IoT networks and wireless networks is shown in Figure 16.2.

16.2 MOTIVATION AND PROBLEM STATEMENT

From the recent studies it has been observed that routing techniques in wireless networks/IoT encounter many challenges as follows: (i) due to failure of some

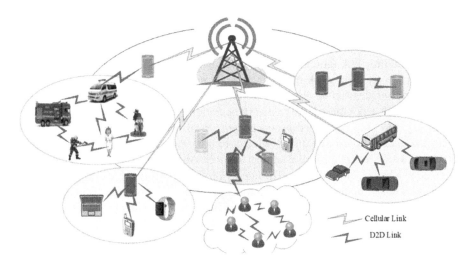

FIGURE 16.1   IoT devices communication example

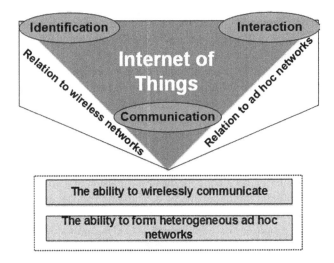

FIGURE 16.2   Relation of IoT and wireless networks

nodes; (ii) protocols are not scalable for large number of nodes; (iii) due to long path communication; (iv) due to jitter variation in the delivery of data; (v) due to single hop communication; (vi) due to load imbalance over some links, etc. These challenges motivated us to design a new protocol for efficient IoT that could take into account multiple parameters. Designing power efficient techniques for IoT networks is very critical and its complexity is amplified as WSN protocols. The major objectives of this chapter are: (1) minimize energy expenditure and (2) link stability estimation based on the traffic induced on each path using the proposed framework for the IoT.

16.3   RELATED WORK

Various routing algorithms for wireless networks have been reported that focus on network stability, link failure among mobile devices, energy constraints, traffic congestion, but very little work has been done pertaining to energy efficiency for IoT. Routing techniques can be classified as: (i) reliability and network operation-based; (ii) energy efficiency-based; (iii) network operation-based.

In [4] Cho et al. proposed a novel hierarchical network architecture that deals with the energy hole issue and devised a routing technique to handle low energy devices. Nodes are placed in a random manner. Moreover, intermediary nodes are located in a hierarchical manner. The energy of intermediary nodes is high compared with other sensor nodes. These intermediary nodes are one hop away from the sensor nodes. This paper stated that the hierarchical intermediate node placement consumes a smaller amount energy.

In [5] authors suggested a Cartesian distance-based transmission technique for the IoT network routing to enhance network lifetime. The routing protocols can only be applied where devices have limited battery resource.

In [6] Necip et al. proposed a routing protocol based on expected transmission count and parent count IoT. The technique used is based on parent devices available and predictable transmission to find the best path. It compared the expected transmission count (ETX) of all parent devices and the parent having least ETX value is selected as the favored parent.

In [7], authors suggested an energy efficient probabilistic routing (EEPR) technique for IoT, which takes into account the residual energy of all sensor nodes in AODV. This technique reduced the number of relay requests to avoid congestion in IoT networks.

In [10], the authors suggested an algorithm for heterogeneous IoT networks energy-harvesting-aware routing algorithm (EHARA). The EHARA technique chooses the optimal route based on residual energy, consumed energy and harvested energy cost parameters of the intermediary node for data transmission toward the sink node. The proposed scheme offered minimum energy consumption and extended the nodes and lifetime of network for mobile dispersed IoT networks.

In [17] authors proposed the improvement LOADng protocol for IoT scenarios. In this paper some improvements have been proposed, first minimum control overhead messages for route discovery, and, second, reducing the number of data packets lost over internet. Various scenarios like sparse and dense networks have been taken by authors and presented significant improvement in terms of energy efficiency and reliability.

In [18], the author presented EMRP – energy aware multi path routing protocol. In this protocol the information regarding remaining battery power is shared with the neighboring nodes in addition to the routing table information. Therefore, when the optimal path is selected, it considered the remaining energy of all the nodes present in the path. In addition to this, nodes also share the queue length size information for predicting the load on particular node. Nodes with a heavy load will deplete their power sooner in comparison with the nodes with fewer loads. Thus, the proposed technique combined remaining energy with the queue length size to get optimal results.

In [19], the author presented a relationship among packet collision in the medium and total energy consumed by the network. The main reason for energy loss is due to channel impairments, where the packets are sent but never reach their destination. Therefore, lost or damaged packets need to be retransmitted by the sender, which leads to energy waste. Thus EECA (energy efficient collision aware) technique is proposed to find collision-free alternative routes and then the packets are transmitted through routes requiring minimum power consumption.

In [20], the author introduced a blocking queuing model in D2D communication to address the issue of fair resource allocation in the network. The author proposed a multi-objective function for minimizing the maximum blocking probability experienced by any type of user in the system.

In contrast, in [21], the author suggested Multi-Path OLSR version 2 to determine multiple disjoint routing paths in the network. It used Dijkstra's shortest path algorithm to discover multiple paths form source to sink node depending on the network topology. During the link failure the proposed algorithm avoid the disjoint route and transmit the data on alternative path to enhance the throughput. The proposed scheme increases the overall reliability and network throughput in highly dynamic conditions.

IoT must support efficient and effective data collection, transmission and data processing in bigger scale as compared to WSN. This paper provides a simple and efficient model to address the energy saving challenges in IoT, which differs from the traditional approaches in the following aspects. First, this chapter considers the energy utilization of both transmitting and receiving data for communications among nodes in IoT. Second, for improving network load, data traffic load estimation of a node is proposed to address the energy depletion issue of some nodes in network. Third, link lifetime is estimated based on the current traffic and alternative route will be selected if energy falls below threshold level.

16.4 MODEL FOR AN ENERGY EFFICIENT IOT

16.4.1 Proposed Work Description

In traditional approaches, the optimal route is selected based viz. energy consumption, link stability, residual energy etc. However, the high energy nodes may lead to link instability as the drain rate can be high of such routes due to frequent selection of the same route again and again. Hence, a new algorithm is proposed in this chapter that takes into account the entire above mentioned factor along with load balancing techniques and avoids the premature death of nodes/networks.

16.4.2 Estimation of Energy Consumption

Power is a limited resource for devices that operate in wireless communiqué for data transmission, forward and reception. Usage of a tremendous quantity of power causes depletion of energy and thus affects overall network performance and routing. Mainly,

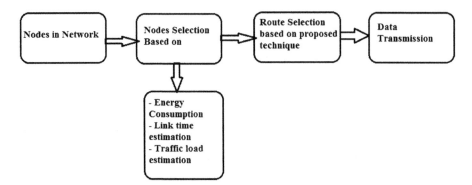

FIGURE 16.3   Framework of proposed protocol

the energy expenditure takes place while transmitting and receiving data packets. Therefore, energy utilization during transmission and reception of data packets can be computed as follows:

$$E_{TX}(c,\mu) = \begin{cases} cE_{TX-elec} + c\varepsilon f_s \mu^2 & \mu < d_0 \\ cE_{TX-elec} + c\varepsilon f_{amp} \mu^4 & \mu > d_0 \end{cases} \tag{1}$$

$$E_{RX}(c) = cE_{RX-elec} \tag{2}$$

Where μ^2 is free space model power loss and (μ^4 *power loss*) is multi-path fading model. These models are used to find out the distance between TX and RX nodes. μ defines the distance between the Tx and Rx. When the distance between Tx and Rx is lower than the threshold value (d_0), the free space model is utilized. Whereas, when the distance between Tx and Rx is higher than d_0, the multi-path fading model is utilized. $cE_{TX-elec}$ and $cE_{RX-elec}$ are defined as the energy consumed per bit by Tx and Rx circuits. Also, the power amplification factors ε_{amp} and εf_s of the and multi-path radio and free space models, respectively.

16.4.3   ESTIMATION OF LINK TIME OF A NODE

The lifetime of a node can be computed by two factors, consumption required by the node and the battery capacity (C_b). A node's lifetime can be calculated as follows:

$$N_{Lt} = \frac{C_b}{Consumption} \tag{3}$$

Energy drain rate[15] can be calculated as follows:

$$DR = \sum e_{ij}^t \sum q_{ij} + \sum e_{ji}^r \sum q_{ji} \tag{4}$$

e^r_{ji} is the reception energy and e^t_{ij} is the transmission energy from node i to node j and q_{ij} is the rate at which traffic flows.

A node's lifetime [14] can be computed as

$$N_{Lt} = \frac{Capacity\ of\ anode}{Drain\ Rate} \tag{5}$$

$$N_{Lt} = \frac{C_{max}}{\sum e^t_{ij} \sum q_{ij} + \sum e^r_{ji} \sum q_{ji}} \tag{6}$$

16.4.4   Data Traffic Load Estimation of a Node

Data traffic load refers to the number of maximum data packets queued at a node. Backpressure algorithms [3] can be used to evaluate the queue length size of a node. $Q^c_x(t)$ denotes the amount of packets flow from a node c ∈ D are backlogged at the node n ∈ D at the time t. Hence, on a link (x,y) of an intermediary device, the backlogged packets can be defined as[16]:

$$Q^c_{(x,y)}(t) = Q^c_x(t) - Q^c_{(y)}(t) \tag{7}$$

Backlogged packets information of all the devices in the network are maintained locally through the information packets exchanged with each other. A weight assigned to the link (x, y) over flow c at the time t is estimated as below:

$$w^c_{(x,y)}(t) = max[Q^c_{(x,y)}(t), 0] \tag{8}$$

Backpressure weight on the link (x, y) is denoted by $w^c_{(x,y)}(t)$ which is the highest weight assigned to the link (x, y) at the time slot t, i.e., $w^c_{(x,y)}(t) = max_{c \in Nw^c_{(x,y)}(t)}$

Multi Metrics Proposed Algorithm

Algorithm: Multi metrics based optimal path selection routing scheme

Input: $E_{TX}, E_{RX},\ N_{Lt}C_b,\ w^c_{(x,y)}(t)$
Output: Optimal route selection

1. Nodes deployed in fixed area randomly
2. Find all optional routes

For i=1 to n

3. Compute energy consumption on each path i to j

4. Calculate link lifetime estimation for each link

$$N_{Lt} = \frac{Capacity\ of\ anode}{Drain\ Rate}$$

5. Calculate traffic load on each path using Eq.(8)

End for loop

6. If $(P_i.energy > P_j.energy)$ then $(P_i$ and P_j is the energy consumption on path i and j respectively)

        If $(P_i.hopCount > P_j.hopCount)$ then
            If $(P_i.Linktime < P_j.Linktime)$ then
                selectRoute(P_j)
          endif
        endif
    endif
else If $(P_i.energy < P_j.energy)$ then $(P_i$ and P_j is the energy consumption on path i and j respectively)
If $(P_i.hopCount < P_j.hopCount)$ then
      If $(P_i.Linktime > P_j.Linktime)$ then
          selectRoute(P_i)
    endif
    endif
endif
else go to step 3

7. Best path is selected based on the values in steps 4, 5, and 6
8. Until the residual energy is more than the threshold value the selected route will be used for communication
9. End

16.5 SIMULATION SETUP

The main purpose of simulation experiments is to evaluate the performance of the proposed techniques and contrast and compare it with both EMBLR and MP-OLSRv2 in terms of network lifetime, throughput and energy consumption in both small and bigger situations.

16.6 RESULTS AND DISCUSSIONS

The outcome obtained from simulation of the proposed multi-metric technique and EMBLR and MP-OLSRv2 routing techniques with different node speeds is shown in Figures 16.5, 16.6 and 16.7 in this section. In the performance evaluation section performance metrics are defined and critical analysis is presented of the obtained simulation results based on throughput, overall energy consumption and network lifetime.

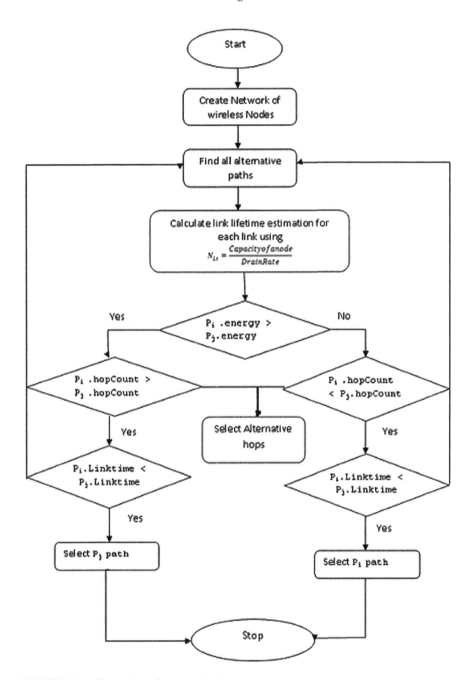

FIGURE 16.4 Flow chart of proposed scheme

TABLE 16.1
Notations

$cE_{TX-elec}$	Energy consumed per bit transmission
$cE_{RX-elec}$	Energy consumed per bit reception
ε_{amp}	Multi-path radio model
εf_s	Free space model
μ^2	Power loss in free space model
N_{Lt}	Node's lifetime
C_b	Battery capacity
$Q_x^c(t)$	Amount of packet flow
$w_{(x,y)}^c(t)$	Backpressure weight of a link
N	Total number of nodes
e_{ji}^r & e_{ji}^t	Transmission energy & reception energy

TABLE 16.2
Parameter settings

Simulation parameters	Value
Node type	TMote sky
Number of sensors	50
Battery capacity	1700 mAh
Transmission range	310 m
Simulation area	400*400
Packet size	512 bytes
Mobility model	RWP model
Simulation time	1,000 sec

The main aim of comprehensive simulation is to evaluate and authenticate the efficiency of the suggested routing technique and analyzed critically QoS parameters:

Throughput: Throughput is defined as the total number of data packets delivered successfully to the destination node in a defined time. It can be calculated as follows:

$$Throughput = \frac{Number\ of\ data\ packets\ sent}{Time} \quad (9)$$

Energy consumption: It defines the average amount of energy consumed in wireless communication during transmission and reception throughout the network simulation time. Energy expenditure of a device can be computed as:

$$Energy\ Consumption = \frac{1}{N} \sum_{i \in N} E_{total}(i) \quad (10)$$

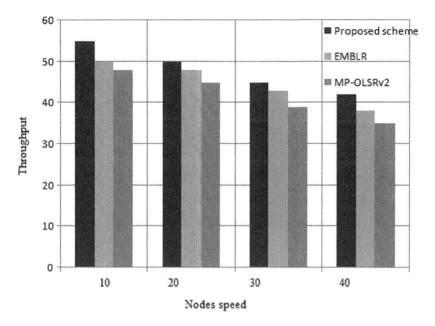

FIGURE 16.5   Nodes speed vs throughput

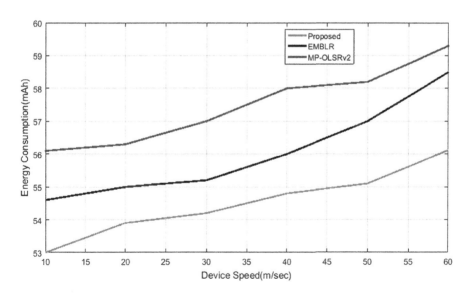

FIGURE 16.6   Device speed vs energy consumption

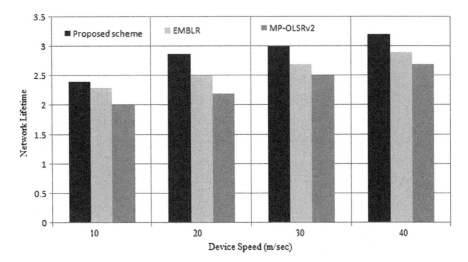

FIGURE 16.7   Device speed vs network lifetime

where $E_{Total}(i)$ is the full amount of energy expended by all nodes in the network and N is the number of nodes in the network.

Network lifetime: Network lifetime is the time when the first node of the network is dead. It creates network partitioning. So we need to choose the optimal routing protocol and use the resources efficiently to increase the network lifetime.

16.7   CONCLUSION

In this chapter, a new multi-metric routing technique is proposed for path selection between devices in IoT networks. It has conquered the challenges of traffic congestion, energy constraints, frequent link failure and network stability of the intermediary nodes for optimal route selection. Hence, the proposed routing scheme exploited link stability, energy consumption and data traffic estimation of an intermediary node in the optimal route selection mechanism to address the challenges if IoT networks. Simulation results clearly show that the proposed technique outperforms compared to EMBLR and MP-OLSRv2. The proposed scheme has shown significant improvement in network lifetime and energy consumption compared to other protocols shown here.

BIBLIOGRAPHY

[1]   Atzori, L., Iera, A., and Morabito, G., The Internet of Things: A survey. *Computer Networks*, 54, 2010, pp. 2787–2805.
[2]   Seth, B., and Dalal S. (2018) Analytical assessment of security mechanisms of cloud environment. In: Saeed, K., Chaki, N., Pati, B., Bakshi, S., and Mohapatra, D. (eds) *Progress in Advanced Computing and Intelligent Engineering. Advances in Intelligent Systems and Computing*, vol 563. Springer, Singapore. https://doi.org/10.1007/978-981-10-6872-0_20

[3]   Bui, L., Srikant, R., and Stolyar, A., Novel architectures and algorithms for delay reduction in back-pressure scheduling and routing. In *Proceedings of the IEEE INFOCOM 2009*, Rio de Janeiro: IEEE, 2009, pp. 2936–2940

[4]   Cho, Y., Woo, S., and Kim, M., Energy efficient IoT based on wireless sensor networks for healthcare. *International Conference on Advanced Communications Technology (ICACT)*, Chuncheon-si Gangwon-do, South Korea:: IEEE 2018.

[5]   Santiago, S., and Arockiam, L., E2TBR: Energy efficient transmission based routing for IoT networks. *International Journal of Computer Science Engineering and Information Technology Research (IJCSEITR)*, 7(4), 2017, pp. 93–100.

[6]   Gozuacik, N., and Oktug, S., Parent-aware routing for IoT networks chapter: Internet of Things, smart spaces, and next generation networks and systems, In *Lecture Notes in Computer Science*, 9247. Cham: Springer, 2015, pp. 23–33.

[7]   Park, S. H., Lee, J. R., and Cho, S., Energy efficient probabilistic routing algorithm for Internet of Things. *Journal of Applied Mathematics*, 1, 2014, pp. 1–7.

[8]   Venckauskas, A., Jusas, N., Kazanavicius, E., and Stuikys, V., An energy efficient protocol for the Internet of Things. *Journal of Electrical Engineering*, 66, 2015, 47–52.

[9]   Dhumane, A., Prasad, R., and Prasad, J. Routing issues in Internet of Things: A survey. *Lecture Notes in Engineering and Computer Science*, 2221, 2016, pp. 404–412.

[10] Nguyen, T. D., Khan, J. Y., and Ngo, D. T., A distributed energy-harvesting-aware routing algorithm for heterogeneous IoT networks. *IEEE Transactions Green Community Network*, 2, 2018, pp. 1115–1127.

[11] Seth, B., and Dalal, S., Analytical assessment of security mechanisms of cloud environment. In Saeed, K., Chaki, N., Pati, B., Bakshi, S., and Mohapatra, D. (eds) *Progress in Advanced Computing and Intelligent Engineering*, Advances in Intelligent Systems and Computing, vol. 563. Singapore: Springer, 2018. https://doi.org/10.1007/978-981-10-6872-0_20

[12] Li, X., Lu, R., Liang, X., Shen, X., Chen, J., & Lin, X. (2011). Smart community: An internet of things application. *IEEE Communications Magazine*, 49(11), 68–75.

[13] Radi, M., Dezfouli, B., Bakar, K. A., & Lee, M. (2012). Multipath routing in wireless sensor networks: Survey and research challenges. *Sensors*, 12(1), 650–685.

[14] Vijarania, M., Jaglan, V., and Mishra, B. K., Modelling of energy efficient algorithm considering temperature effect on lifetime of a node in wireless network. *International Journal of Grid and High-Performance Computing*, 12(2), 2020, pp. 87–101.

[15] Habib, S., Saleem, S., and Saqib, K. M. Review on MANET routing protocols and challenges. *Proceedings of the IEEE Student Conference on Research and Development, Putrajaya, Malaysia*, 12, 2013, pp. 529–533).

[16] Seth, B., Dalal, S., and Kumar, R. Hybrid homomorphic encryption scheme for secure cloud data storage. In Kumar, R., and Wiil, U. (eds) *Recent Advances in Computational Intelligence*, Studies in Computational Intelligence, vol. 823. Cham: Springer, 2019. https://doi.org/10.1007/978-3-030-12500-4_5

[17] Sobral, J. V. V., Rodrigues, J. J. P. C., Rabelo, R. A.L., Saleem, K., and Furtado, V., LOADng-IoT: An enhanced routing protocol for Internet of Things applications over low power networks. Sensors, 19(150), 2019, pp. 1–26.

[18] Li, M., Zhang, L., Li, V. O., Shan, X., and Ren, Y., An energy-aware multipath routing protocol for mobile ad hoc networks. *ACM Sigcomm Asia*, 5, 2005, pp. 10–12.

[19] Wang, Z., Bulut, E., and Szymanski, B. K., Energy efficient collision aware multipath routing for wireless sensor networks. In *Proceedings of the 2009 IEEE International Conference on Communications*, Dresden: IEEE, 2009, pp. 1–5.

[20] Katsinis, G., Tsiropoulou, E. E., and Papavassiliou, S., *On the Problem of Resource Allocation and System Capacity Evaluation via a Blocking Queuing Model in D2D Enabled Overlay Cellular Networks*. Berlin/Heidelberg: Springer,, 2015, pp. 76–89.

[21] Parrein, B, and Yi, J., *Multipath Extension for the Optimized Link State Routing Protocol Version 2 (OLSRv2)*. Reston, VA: Internet Society, 2017.

17 Green Cloud Computing

A Step Towards Environment Sustainability Using Live Virtual Machine Migration

N. Mehra, D. Kapil, A. Bansal, and N. Punera

CONTENTS

17.1 INTRODUCTION

In the last two decades, sustainability has gained importance amongst developers (both hardware and software developers) and users, due to the speedy progress in consumption of energy. In order to elevate green along with environment friendly sustainable projects, i.e. ICTs, the impact of ICTs on the environment during the complete steps of process has been calculated. The need to conserve energy along with information in cloud-based ICT solutions is at present becoming a serious issue in this field. Cloud computing is a most suitable expertise to provide a variety of information technology services intended for organizations that facilitate different on-demand pay-for-use virtual resources [1, 2]. As there is a raise in internet-based requirements or cloud services, companies are investing in building large data center to host cloud services. NIST [5] define cloud computing as:

> Cloud computing is a model for enabling ubiquitous, convenient, on-demand network access to a shared pool of configurable computing resources (e.g., networks, servers, storage, applications, and services) that can be rapidly on and released with minimal management effort or service provider interaction.

However, not only are these data centers complex to build but they also use a large amount of energy. So there is a pressure on the manufacturers to develop methods, products and services that lessen the adverse consequence on the ecosystem. Therefore, technologies such as virtualization and consolidation are used to make these data centers energy efficient.

17.1.1 CARBON FOOTPRINT CHALLENGES AND IN CLOUD ENERGY CONSUMPTION

Cloud services are delivered through huge data centers (DCs) that consist of several servers and physical machines (PMs) are utilized by cloud service providers (CSPs) [3, 4]. One of the foremost concerns with these large-scale data centers is that they require a huge quantity of energy, which mostly comes from fossil fuels. Based on a report from NRDC [6], US DCs used 91 billion kWh of electricity only in 2013. With the regular growth of the economy all around the world, fossil fuel-generated energy consumption will grow in future. According to R. Kumar and L. Mieritz [8], every year ICT-based CO_2 emissions are rising 6% per year, and by means of such a rapid rate of growth, they also forecast they will comprise approximately 12% of global emissions by 2020. By making energy-efficient DCs, cloud service providers are capable to decrease energy usage and CO_2 emissions.

To conquer the trouble of energy consumption with ecological concerns due to the large amount of CO_2 discharge from energy sources, some feasible solutions can be improving the working DCs. By building data centers that are power competent as well as sensitive towards power sources, cloud service providers are capable to trim down the energy use or usage and carbon footprint significantly [9].

17.1.2 ENVIRONMENT FRIENDLY CLOUD COMPUTING: GREEN COMPUTING

The origins of green computing were in 1987, following the World Commission report name "Our Common Future." It is fundamentally focused on sustainable development [10,11]. Liu et al.[12] anticipated green cloud construction, which was designed to decrease power consumption in data centers without affecting the cloud characteristics. Figure 17.1 gives the architectural view of green cloud computing and Figure 17.2 illustrates categorization of energy and carbon level techniques, i.e. green cloud techniques.

17.2 BACKGROUND

Nowadays, most organizations are working to improve the energy effectiveness of cloud data centers. Recent research concentrates on issues related to green ICT and reducing energy in today's cloud computing systems.

Webster and Watson [17] stated that literature reviewing is a fundamental step towards advancing knowledge on any topic. Researchers [13, 28] have conducted research in reduction of carbon footprints in cloud computing technology and recapitulating taxonomy of green cloud computing studies.

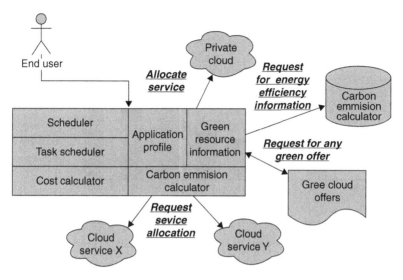

FIGURE 17.1   Energy efficient cloud architecture, i.e. green cloud

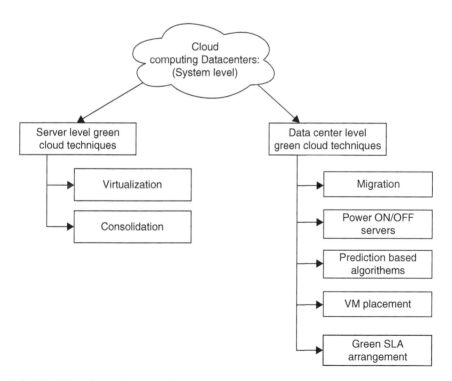

FIGURE 17.2   Categorization of energy and carbon level techniques, i.e. green cloud techniques

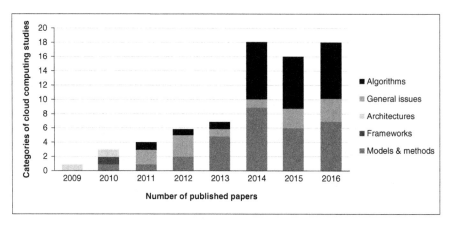

FIGURE 17.3 Literature review based on five categories of energy efficient cloud computing [28]

Baliga et al. [35] have written that with the growing trend of cloud computing, there is a need for network-based computing over traditional office-based space of computing. With the overall change in the concept of computing, there is a need to manage the energy consumption in the field of complete ICT (information and communication technology). More attention has been given to energy use in data centers and less to the key components that are responsible for connecting users to the cloud, which includes switching and transmission networks. The work in this chapter focuses on energy consumption of cloud computing in both the public and private sector and finds that the energy consumption in transmission and switching makes a significant percentage of the total energy consumed.

Avgerinou et al. [36] have focused on a major challenge that human beings are going through, which is climate change. The use of various technologies in ICT is leading to a significant percentage contribution in the total carbon dioxide the world emits. The technologies that are a major part of it are cloud computing and the use of the internet all over the world. The articles talk about the recent trends in the consumption of energy and evaluating the efficiency of the Europe Union data centers. An energy efficiency program was created in 2008 to address the increasing energy consumption in these data centers and the impact was measured.

Dougherty et al. [37] discuss that with the use of cloud computing's virtualization, power consumption can be reduced by the use on demand feature. With the help of auto-scaling, we can use resources precisely according to our needs and the problem of allocating a large amount the wastage is also eliminated. A model is suggested that takes into account the auto-scaling feature and takes care of the configuration, consumption of energy and also the cost so as to provide a better, greener computing environment and eliminates the emission from the idle resources.

According to Jain et al. [38], the evolution of cloud computing has brought about a huge demand for the high-performance data centers, and with the growth in the use of these data centers the emission has also increased tremendously. The demand of green computing arises with the high amount of carbon dioxide that is emitted during the use of these technologies. Green computing suggests saving energy, recycling and also minimizing wastage. High-speed computing means more processors/chips, the more chips the more heat they produce, and more heat needs more cooling, which again emits heat, so we need a method that provides high speed with less heat.

Chen et al. [39] discuss the high-performance computing infrastructure in Netherlands using cloud. Here the discussion is on three features, which are the power metric, power efficiency metric and lastly the energy metric, which are taken account of under high-performance workloads. A linear power model is created where the behavior is represented as a single work node and has the contribution of several working components such as CPU, HDD and memory. The result shows that a power characterization module can be integrated into cluster monitoring and green cloud computing can have a huge impact on the energy consumption and reuse.

Mohsenian-Rad and Leon-Garcia [40] discuss the fact that the increase in the use of internet has brought about the need for high performance data centers, which are high performing but at the same time consume a lot of energy. With the high demand of performance, they cause a huge impact of power grid, and emit a high amount of greenhouse gases, which has a major impact on our environment. The paper presents an optimization framework, which aims to minimize the energy cost and also minimize the carbon footprints and also retain the quality which has to be provided.

Masdarin and Zangakani [29] surveyed the virtual machine placement method, published between 2010 and 2019, and analyzed that the graph is increasing consistently. They focus their study in different search strings such as predictive VM survey, review and placement. Their studies also reflect that effective VM placement is one of the crucial issue of cloud computing, which directly affects the cost, scalability and most importantly energy usage in cloud computing, which is directly related to the CO_2 emission, which is directly related to the environment sustainability. The article also provides a systematic survey on the below parameters:

- Power usage
- Memory usage
- Workload usage
- Resource usage
- Resource demand
- Cost
- Demand consumption
- CPU usage
- Resource requirements
- Temperature
- Maximum sustainable throughput
- Network load
- Number of VM.

Barroso and Hölzle [4] provide the real-life usage of energy-proportional designs in large energy savings in servers. In their study they compare dollars with CO_2 and laptops with servers. They highlighted that the mismatch between the problem of divergence between the servers' energy-efficiency individuality and the performance of server-class workloads is the prime responsibility of designers.

Kolodziej et al. [13] emphasize the use of live virtual migration in memory utilization in memory intensive applications in the cloud environment and also develop metrics for calculating different types of cost in cloud environments. In their paper they also given pseudo code for main call-back for LVMM implement in QEMU-KVM. They also present an algorithm for achieving low downtime and low application performance impact.

Buyya et al. [14] proposed a power consumption model, which identifies a correlation between the CPU usage time, CPU work load and total CPU energy.

Bala and Green [15] suggested the significance of VM relocation in cloud DCs, which is primarily focused on energy performance. They present an automated system named Sandpiper for detecting hotspots and shaping mapping of physical to virtual resources and also capable of initiating necessary required migration. They implement their technique in Xen.

Anton et al. [16] proposed a framework based on energy-efficient green cloud computing. They also define energy aware methods to utilize data center resources, focusing on delivery of negotiable quality of service.

Liu and Fan [33] presented their findings based on CPU scheduling for transparent live migration, in which the CPU scheduling mechanism is based on target host executable host files used for adjustable log generation rate.

Xu et al. [42] emphasize the problem of VM placement in data centers in respect of power consumption, resource wastage and maintaining the temperature of the servers. In their paper they use a genetic algorithm to perform the process of VM placement in data centers. They use a genetic algorithm on the global controller of data center to perform VM placement. The worldwide controller gets the VM demands and afterward, dependent on a multi-objective VM arrangement calculation, relegates each VM to a server. This calculation, the same as the recently examined work, performs the VM position in the wake of accepting all the VM demands, which isn't a unique way. In addition, the calculation balances between power utilization and temperature.

Haque el al. [43] suggest a novel class of cloud services that provides a specific service stage agreement to the user to achieve the required environment friendly energy to run their applications. Their study is based on the solar energy-based method proposed by I. Goiri et al. [44].

Haque el al. [43] also embrace another force foundation where each rack can be fueled from earthy colored or environmentally friendly power vitality sources. The streamlining strategies have the target of expanding the supplier's benefit by conceding the approaching occupations, with green SLA prerequisites. In the event that the cloud supplier can't meet the level of efficient power to run, the activity should take care of inconveniences for the client.

The Google Compute Engine [17] has used the concept of live virtual machine migration since 2013 to keep customers' virtual machines running while fixing H/W problems, updating S/Ws or tackling unexpected issues. Table 17.1 gives a summary of methods, objectives and outcomes given by different researches.

TABLE 17.1
Summary of methods, objectives, and outcomes given by different researchers

S. No	Ref	Method	Objective	Outcome
1	[18]	Energy effective methods in cloud computing	Define metrics for optimization of H/W resources and minimization of power consumption	If we minimize migration time, QOS improved
2	[19]	Virtual machine Migration techniques	Providing method for minimize relocation i.e. migration and implement DVFS by using. Net framework	Proposed metrics based on Energy usage, SLA disobedience and no. VM relocation. The explore SLA violation with 40% intermission amongst thresholds.
3	[20]	Task scheduling	Reduce carbon footprints	Consider two algorithms: 1. Dynamic cloud min-min scheduling 2. Dynamic cloud list scheduling, verified their findings using CLOUD SIM simulator
4	[21]	Based on detection of memory update patterns	Enhanced Pre-Copy approach based on frequent updating of memory pages memory.	It is very useful for the memory exhaustive applications
5	[22]	Enhanced Pre-Copy VM relocation approach	Time series forecasting method	VM migration process which is useful when iteration consist of high dirty pages
6	[23]	Green Slot scheduler	Make best use of green energy usage and decrease cost in terms of brown energies	Based on Prediction technique for the best use of solar energy and hang up the set jobs if green energy is not available
7	[24]	Migration of VM using Post-copy Approach	Reducing migration time	In migration the entire virtual machine migration process all the memory pages to destination server dispatched only once.
8	[25]	VM placement using minimum migration policy	Optimization of VMs	If VM is consolidated it gives minimization of carbon footprints
9	[30]	Virtual machine Migration techniques	Based on the coping pages to targeted Virtual Machine from identified	More refined Stop-and-Copy approach of virtual machine migration

17.3   SERVER CONSOLIDATION FOR ENERGY EFFICIENCY

In the field of energy efficient cloud computing, server consolidation can be defined as a systematic process/ technique to make use of a dedicated physical server for user instances or server applications. The main focus of this technique is to reduce the energy usage in an organization by reducing the total number of active servers. It is also considered as one of the useful techniques in migrating a small load of different computers/servers to a single server/computer.

17.3.1   NEED OF VIRTUALIZATION

Energy usage of cloud data centers is directly associated with the processing power essentials and also affected by the active no of physical host. A. Corradi [30] defines the consolidation of dynamic virtual machines for minimizing energy consumption in the cloud. According to K. Divya [31]: "Virtualization can be applied to either single resource or to a complete computing system, also known as Platform Virtualization; it allows multiple 'Virtual Machines' into the same computing host."

Just imagine a scenario of peak operational hours of peak working days. How many total requirements will be requested of the cloud data centers? Furthermore, how many total "virtual machines" will be required to run over hosts and physical machines? Although the late night hours and on holidays scenario is total different, fewer "virtual machines" need to be allocated on physical machines in cloud data centers. Therefore, we can simply decrease the energy consumption in the physical host by migrating and consolidating the VMs in DCs. Virtualization can be classified in two basic categories; the taxonomy is given in Figure 17.4

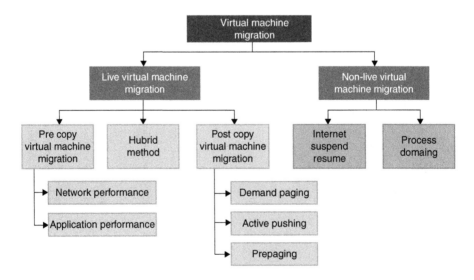

FIGURE 17.4   Virtual machine migration

17.3.2 Methodology: Energy Efficiency Using LVMM (Live VM Migration)

Energy consumption and utilization of the resources in data centers are significant features in cloud computing. Due to virtualization technology the resources of the datacenters are used effectively using virtual machines. Virtualization allows a physical machine to run many virtual machines, a virtual machine is the abstraction of the physical machine. Virtualization software such as Xen, KVM, VMWare, Virtual box, etc., called a virtual machine monitor or hypervisor, is used to make a virtual environment. Virtualization provides several benefits like isolation between virtual machine to virtual machine, better resource utilization, reliability, fault tolerance. Live virtual machine migration is a very important feature provided by virtual machine monitors, which provides several advantages such as load balancing, server consolidation, availability of the services when the machine is under maintenance these services which are running in the virtual machines, can be migrated to other physical machine. Energy consumption can be reduced using live virtual machine migration as some physical machines are underutilized, then these physical machines can be shut down and their virtual machines can be migrated to other physical servers. The consolidation of servers provides the energy efficient green cloud. Live virtual machine migration can be categorized as cold (non-live) and hot (live). In cold migration, the virtual machine is paused, and it is noticeable by the customer whereas in hot migration there is no service interruption. Further, live virtual machine migration techniques can be categorized into post copy and pre copy. In the post copy approach, first the virtual machine that is to be migrated at the source host is suspended and the CPU's minimum state is copied to the destination host and the virtual machine is resumed and memory pages are fetched over the network. While in the pre-copy approach, two steps are followed. The first step is warm up and the second step is stop and copy. In the first step , i.e. warm-up phase, the hypervisor copies all the memory pages from the source host and some pages that change in the migration process duration are marked dirty pages and later these pages are re-sent until the given threshold. Stop-copy step stops the virtual machine at the source host and the remaining pages are resent and the virtual machine is resumed at the destination host.

LVMM is the practice that refers to the migrating resources (CPU, storage, memory) and application as of one physical server (source machine) to a new/another physical server (destination server). This technique is very useful in decreasing energy usage in cloud computing. Some of the features of live VM migration are given below in Figure 17.5.

To reduce energy consumption we present a scenario given below in Figure 17.5 in which we present a data center, which has three physical servers. In the first physical server (PS_1) we have three VMs, i.e. VM_1, VM_2, VM_3, a second physical server (PM_2), which has a light workload and runs only one VM, i.e. VM_4, and a third physical server (PM_3) that has two physical machines, VM_5, VM_6, for saving energy consumption, when the workload is less, for example late night hours, weekends and on holidays. To save energy when the workload is less, we are migrating VM in physical server 2 to physical server 3 by using Live Virtual Migration (LVMM). By migrating

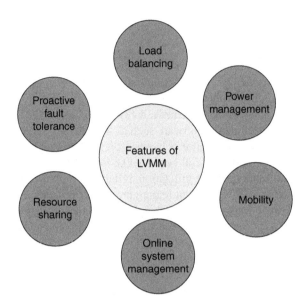

FIGURE 17.5 Features of live virtual machine migration

VM$_4$ to the third physical server, all the work of PS$_2$ is migrated to PS$_3$ and we are able to decrease energy consumption, i.e. energy efficient cloud computing.

The present scenario is based on the technique of LVMM, which is based on the Goggle Compute Engine VM migration [17].

The three-phased live migration process consists of these steps [17]:

1. Start the migration process.
2. Initialize:
   a. The VM to migrate from identified source.
   b. Server selects the destination server.
3. Pre-copy the virtual memory of the selected VM to the target server at the same time as the VM is still operating on the source server and assume it is dirty.
4. Repeat step 3 until all dirty pages are copied to the starting server VM to target server.
5. Suspend VM and copy state to target server; send source virtual machine to non-memory state.
6. Resume VM at destination server.
7. Stop the migration.

Figure 17.7 will show the flow chart representation of mentioned scenario given in Figure 17.6.

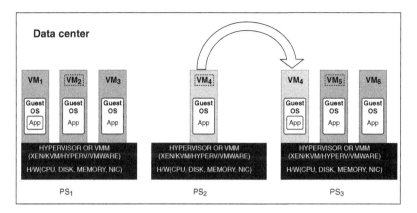

FIGURE 17.6 Energy efficiency using live VM migration

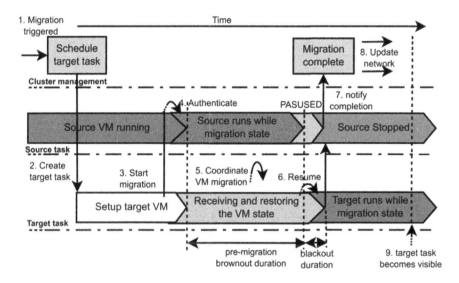

FIGURE 17.7 Google Compute Engine VMM [17]

17.4 CONCLUSION

In today's internet-based online world, outgrowing cloud-based demands leads to increasing the number of data centers, which leads to higher demand for more energy. Energy consumed by data centers (DCs) is one of the attentive concerns in many conferences, as environment sustainability is a key concern of entire world. The more energy in DCs leads to a huge amount of CO_2 emission and as a result increased carbon footprints, as DCs need a huge amount of energy, which usually comes from fossil fuels because it is very difficult to generate such a huge amount energy from green or renewable sources. Here the discussion is about the need for green cloud

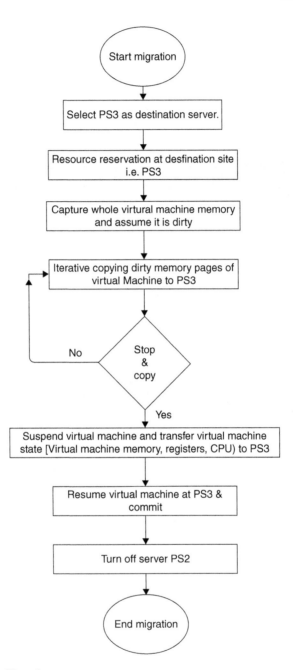

FIGURE 17.8 Flow chart

computing and procedures. We present a case study scenario. This chapter aims to identify a technique to reduce energy usage in DCs. The significance of this chapter is to present pre-copy Live Virtual Machine Migration (LVMM) as a solution to achieve green cloud computing by reducing energy usage in cloud DCs. In future, we will implement this scenario in a green computing simulator.

BIBLIOGRAPHY

[1] M. Alizadeh, S. Abolfazli, M. Zamani, S. Baaaharun, and K. Sakurai, "Authentication in mobile cloud computing: A survey," *Journal of Network. Computer Applications*, 61, pp. 59–80, 2016.

[2] M. Masdari, F. Salehi, M. Jalali, and M. Bidaki, "A survey of PSO-based scheduling algorithms in cloud computing," *Journal of Network Systems Management*, 25(1), pp. 122–158, 2017.

[3] M. Nazari Cheraghlou, A. Khadem-Zadeh, and M. Haghparast, "A survey of fault tolerance architecture in cloud computing," *Journal of Network Computer Applications*, 61, pp. 81–92, 2016.

[4] J. Varia, "Best practices in architecting cloud applications in the AWS cloud," In Buyya, R., Broberg, J., Goscinski, A.M. (eds) *Cloud and Computing: Principles & Parag*, Hoboken, NJ: Wiley, pp. 457–490, 2011.

[5] P. Mell and T. Grance, "The NIST definition of cloud computing," Version 15, 10-7-09, www.wheresmyserver.co.nz/storage/ media/faq-files/cloud-def-v15.pdf.

[6] B. Seth, S. Dalal, and R. Kumar, "Hybrid homomorphic encryption scheme for secure cloud data storage." In Kumar, R., and Wiil, U. (eds) *Recent Advances in Computational Intelligence*, Studies in Computational Intelligence, vol. 823. Cham: Springer, 2019. https://doi.org/10.1007/978-3-030-12500-4_5

[7] E. P. F. Lee, J. Lozeille, P. Soldán, S. Daire, J. Dyke, & T. Wright, "An ab initio study of RbO, CsO and FrO (X2Σ+; A2Π) and their cations (X3Σ-; A3Π)," *Phys. Chem. Chem. Phys.*, 3(22), pp. 4863–4869, 2001.

[8] R. Kumar and L. Mieritz, *Conceptualizing Green IT and Data Centre Power and Cooling Issues*. Gartner Research Paper No. G00150322,2007. www.gartner.com.

[9] J. W. Smith and I. Sommerville, *Green Cloud: A Literature Review of Energy-Aware Computing*, St Andrews: Depd Sys. Engg Group School of Comp. Science, Uni. of St Andrews, , 2010.

[10] B. Guo, Y. Shen, and Z.-L. Shao, "The redefinition and some discussion of green computing,"*Jisuanji Xuebao/Chinese Journal of Computers*, 32, pp. 2311–2319, 2009.

[11] Z. Juntao, *Urban and Regional Sustainable Development*, Dongbei: Dongbei University of Finance and Economics Press, 2008.

[12] L. Liu, H. Wang, X. Liu, X. Jin, W. Bo He, Q. B. Wang, and Y. Chen, "GreenCloud: A new architecture for green data center," *CAC-INDST '09: Proceedings Of The 6th International Conference Industry Session On Autonomic Computing And Communications Industry Session*, June 2009, pp. 29–38, 2009. https://doi.org/10.1145/1555312.1555319.

[13] K. Z. Ibrahim, S. Hofmeyr, C. Iancu, and E. Roman, "Optimized pre-copy live migration for memory intensive applications," *Proceedings of the Int Conf of High Performance Computing, Networking, Storage and Analysis (SC 2011)*, Seattle, WA: IEEE, pp. 1–11, 2011.

[14] R. Buyya, C. Shin Yeo, S. Venugopal, J. Broberg, and I. Brandic," Cloud computing and emerging IT platforms: Vision, hype, and reality for delivering computing as the 5th utility," *Future Generation Computer Systems*, 25(6), pp. 599–616, 2009.

[15] T. Wood, P. Shenoy, A. Venkataramani, and M. Yousif. "Black-box and gray-box strategies for virtual machine migration," in *Proceedings of the 4th USENIX conference on Networked systems design & implementation (NSDI'07),* USENIX Association, 2007.

[16] A. Beloglazov, J. Abawajy, and R. Buyya, "Energy-aware resource allocation heuristics for efficient management of data centers for cloud computing," *Future Generation Computer Systems,* 28(5), pp. 755–768, 2012.

[17] Google Cloud Platform Blog: "Google Compute Engine uses live migration technology to service infrastructure without application downtime." https://cloudplatform. googleblog.com/2015/03/Google-Compute-Engine-uses-Live-Migration-technology-to-service-infrastructure-without-application-downtime.html.

[18] T. K. Okada, A. De La Fuente Vigliotti, D. M. Batista, and A. Goldman vel Lejbman, "Consolidation of VMs to improve energy efficiency in cloud computing environments," *XXXIII Brazilian Symposium on Computer Networks and Distributed Systems,* Vitoria, Brazil, pp. 150–158, 2015.

[19] C. P. Sapuntzakis, R. Chandra, B. Pfaff, J. Chow, M. S. Lam, and M. Rosenblum, "Optimizing the migration of virtual computers," *SIGOPS Operations System Review,* 36, pp. 377–390, 2003.

[20] A. Beloglazov and R. Buyya, "Energy efficient resource management in virtualized cloud data centers," in *10th IEEE/ACM International Conference on Cluster, Cloud and Grid Computing,* Melbourne: ACM, pp. 826–831, 2010.

[21] J. Li, M. Qiu, Z. Ming, G. Quan, X. Qin, and Z. Gu, "Online optimization for scheduling preemptable tasks on IaaS cloud systems," *Journal of Parallel and Distributed Computing,* 72(5), pp. 666–677, 2012.

[22] H. Jin, L. Deng, S. Wu, X. Shi, and X. Pan, "Live virtual machine migration with adaptive, memory compression," in *IEEE International Conference on Cluster Computing and Workshops,* New Orleans, LA: IEEE, pp. 1–10,2009.

[23] B. Hu, Z. Lei, Y. Lei, D. Xu, and J. Li, "A time-series based precopy approach for live migration of virtual machines," in *IEEE 17th International Conference on Parallel and Distributed Systems,* Tainan: IEEE, pp. 947–952, 2011.

[24] Í. Goiri, R. Beauchea, K. Le, T.D. Nguyen, Md. E. Haque, J. Guitart, J. Torres, & R. Bianchini, "GreenSlot: Scheduling energy consumption in green datacenters," in *SC '11: Proceedings of Int. Confn for High Perf. Comp, Netwk., Storg & Analy.,* Seattle, WA, pp. 1–11, 2011.

[25] M.R. Hines, U. Deshpande, and K. Gopalan, "Post-copy live migration of virtual machines," *SIGOPS Operations Systems Review,* 43(3), pp. 14–26, 2009.

[26] J. Kołodziej, S. Ullah Khan, L. Wang, M. Kisiel-Dorohinicki, S. A. Madani, E. Niewiadomska-Szynkiewicz, A. Y. Zomaya, and C.-Z. Xu, "Security, energy, and performance-aware resource allocation mechanisms for computational grids," *Future Generation Compute Systems,* 31, pp.77–92, 2014.

[27] C. P. Napuntzakis, R. Chandra, B. A. Pfaff, J. Chow, M. S. Lam, and M. Rosenblum, "Optimizing the migration of virtual computers," *SIGOPS Operations Systems Review,* 36, pp.377–390, 2003.

[28] L.D. Radu, "Green cloud computing: A literature survey," *Symmetry,* 9(12), p. 295, 2017.

[29] M. Masdari, and M. Zangakani, "Green cloud computing using proactive virtual machine placement: Challenges and issues," *Journal Grid Computing,* pp. 129–134, 2019.

[30] A. Corradi, M. Fanelli, and L. Foschini, "Increasing cloud power efficiency through consolidation techniques," in *IEEE ISCC,* Kerkyra: IEEE, pp. 129–134,2011.

[31] D. Kapil, E. S. Pilli, and R. C. Joshi, "Live virtual machine migration techniques: Survey and research challenges," in *3rd IEEE IACC,* Ghaziabad: IEEE, pp. 963–969, 2013.

[32] M. Nelson, B. H. Lim, and G. Hutchins, "Fast transparent migration for virtual machines," in *Proceedings of the Annual Conference on USENIX Annual Technical Conference (ATEC '05),* USENIX Association, p. 25, 2005.

[33] W. Liu and T. Fan, "Live migration of virtual machine based on recovering system and CPU scheduling,"*6th IEEE Joint Intrl. Info. Tech. and A.I Conf.,* Chongqing: IEEE, pp. 303–307, 2011

[34] M. Avgerinou, P. Bertoldi, and L. Castellazzi. "Trends in data centre energy consumption under the European Code of Conduct for Data Centre Energy Efficiency," *Energies,* 10(10), 2017. https://doi.org/10.3390/en10101470.

[35] J. Baliga, R. W. A. Ayre, K. Hinton, and R. S. Tucker, "Green cloud computing: Balancing energy in processing, storage, and transport," *Proceedings of the IEEE,* 99(1), pp. 149–167, 2011. https://doi.org/10.1109/JPROC.2010.2060451.

[36] Q. Chen, P. Grosso, K. Van Der Veldt, C. DeLaat, R. Hofman, and H. Bal, "Profiling energy consumption of VMs for green cloud computing," In *Proceedings – IEEE 9th International Conference on Dependable, Autonomic and Secure Computing, DASC,* Kerkyra, Greece: IEEE, pp. 768–775, 2011. https://doi.org/10.1109/DASC.2011.131.

[37] B. Dougherty, J. White, and D. C. Schmidt, "Model-driven auto-scaling of green cloud computing infrastructure," *Future Generation Computer Systems,* 28(2), pp. 371–378, 2012. https://doi.org/10.1016/j.future.2011.05.009.

[38] A. Jain, M. Mishra, S. K. Peddoju, and N. Jain, "Energy efficient computing: Green cloud computing," in *2013 International Conference on Energy Efficient Technologies for Sustainability, ICEETS,* Nagercoil, pp. 978–982, 2013. https://doi.org/10.1109/ICEETS.2013.6533519.

[39] A.-H. Mohsenian-Rad and A. Leon-Garcia, "Energy-information transmission tradeoff in green cloud computing," *Carbon,* 100, 2010.

[40] L. A. Barroso and U. Hölzle, "The case for energy-proportional computing," *Computer,* 40(12), pp. 33–37, December2007. doi: 10.1109/MC.2007.443.

[41] J. Xu and J. A. Fortes, "Multi-objective virtual machine placement in virtualized data center environments," in *Proceedings of the Green Computing and Communications (GreenCom), 2010 IEEE/ACM Int'l Conference on & Int'l Conference on Cyber, Physical and Social Computing (CPSCom),* IEEE, pp. 179–188, 2010.

[42] M. E. Haque, K. Le, I. Goiri, R. Bianchini, and T. D. Nguyen, "Providing green SLAS in high performance computing clouds," in *Proceedings of the International Green Computing Conference (IGCC),* Arlington, VA: IEEE, pp. 1–11, 2013.

[43] I. Goiri, K. Le, T. D. Nguyen, J. Guitart, J. Torres, and R. Bianchini, "Greenhadoop: Leveraging green energy in data-processing frameworks," in *Proceedings of the 7th ACM European Conference on Computer Systems,* ACM, pp. 57–70, 2012. https://doi.org/10.1145/2168836.2168843.

18 Smart Education Technology

Design Research of Future Formal Learning Environment in Smart Cities

Jyoti Chauhan and Puneet Goswami

CONTENTS

18.1 INTRODUCTION

Technology has changed many aspects of the education system. Education is the way to transform the way we lead our life. With the initiation of the use of Internet of Things (IoT) in the education system we can keep track of overall resources pertaining to education. IoT not only plays a crucial role in teaching and learning process but also in assessment. From KG to PG, IoT is becoming the basic requirement in all aspects of education. The intensive advancement in technology has changed the learning environment a lot in the past few years (whether inside or outside the classroom).

In this chapter, we describe how education system needs can be better addressed adopting technology named Internet of Things (IoT). This novel technology IoT is referred to as an interconnection of networked devices keeping three aspect in focus (i.e. anytime, anywhere, anyone) and has played an important role in the education industry. It is also called a game-changer by some stakeholders of the education market.

In general, it can be said that technology along with web availability have changed the methodology of how students were taught and students learn in universities. As per a report by Markets and Markets Research Pt. Ltd., worldwide IoT in the education market is expected to increase from USD 4.8 billion in 2018 to 11.3 billion by 2023 at a CAGR of 18.8%.

IoT has the potential to transform the lives of all the stakeholders of the education market (i.e. educators, management, parents, and students). So we can say it has the ability to change a static classroom to a dynamic one (i.e. "anytime, anywhere and anyone") through remote access from connected devices (smart devices like smart phones, applications). It helps in automatic execution and decision-making along with providing some security features. It is also considered a key part in increasing the quality of education.

18.1.1 Smart Classroom and Its Needs

Smart classrooms target improving the learning capacity of students. Smart classrooms and e-learning help in making entire subjects more interesting thus improving students' results. Smart classrooms and e-learning help students of all ages and encourage greater comprehension. Smart classrooms copy every one of the capacities found in real classrooms and can also be viewed as virtual classrooms.

18.1.2 Highlights of Smart Classrooms

There are various highlights of a smart classroom. Not all of them are referenced underneath:

1. **Collaborative and adaptive learning:** Learning in groups creates basic reasoning and improves learning extension. Collaborative learning exercises incorporate discussions and group projects, and that's only the tip of the iceberg. Versatile learning gives students the opportunity to learn at their pace in a manner they are comfortable with.
2. **Execution base evaluation:** Appraisal techniques are not limited to tests. Different tests and different exercises are used to assess students in a superior manner.
3. **Most recent innovations**: Students can utilize utilities that have supplanted conventional paper and pen alongside other gadgets introduced for the improvement of the learning process. Educators find support and the chance to improve the subjects being instructed during a smart classroom session.

4. **Learning management system:** Students will comprehend and accept the techniques and will assume the responsibility for their own learning. Smart classroom are structured as student centric and focused on every student's needs.

5. **New age instruction:** Updated data and overhauled training will be given to students. Traditional training–learning techniques will be followed yet combined with other imaginative strategies, which makes smart classrooms intriguing and captivating.

18.2 WHAT IS THE INTERNET OF THINGS?

The term "Internet of Things" was proposed by Kevin Ashton in 1998 [1] who concluded that "The Internet of Things can possibly change the world, similarly as the Internet did. Perhaps more." At that point the MIT Auto-ID focus presented their IoT vision in 2001 [2]. The IoT expression was officially utilized by the International Telecommunication Union (ITU) in 2005 [3].

18.2.1 ELEMENTS OF IoT

The components of IoT are identification, sensing, communication, computation, services, and semantics as depicted in Figure 18.1.

All in all, there are four primary segments in IoT development:

- **The Internet:** which provides communication between everything whenever, anyplace.
- **The hardware:** the embedded communication hardware, for example, sensors, tags, actuators, and transceivers.
- **The middleware:** for information storage space, processing, and perspective understanding.
- **The presentation:** for representation along with translation equipment for various stages.

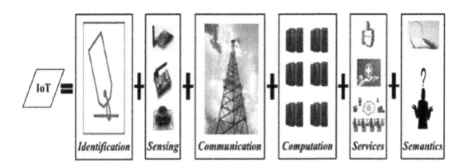

FIGURE 18.1 Elements of IoT

Identification techniques, for example, electronic product code (EPC) plays an important role in the identification of articles and sending its data to databases and information stockrooms such as data warehouses and so on, which is then examined to perform explicit activities dependent on required services. IoT services include the identification services, which empower recognizable proof of smart gadgets in the real world. The collaborative related services follow up on crude information. Pervasive services provide ongoing activity on request.

18.2.2   FOCUS

The aim of introducing this novel innovation in the education sector was to:

• Individualization of learning
• Interactivity
• Homogeneous quality
• Quick conveyance of data
• Reduction in the cost of education

18.2.3   FACTORS PROMPTING ACCOMPLISHMENT OF IoT IN EDUCATION SECTOR

Key elements for fruitful usage of IoT in education are listed below:

1.   **Anytime and anyplace learning:** learning isn't reliant on a specific location and time
2.   **Security:** providing a secure learning environment
3.   **Innovative, intuitive, integrative, community and evaluative learning**
4.   **Instructive policies**

18.2.4   HOW IS THE EDUCATION SECTOR USING IoT FOR PROGRESS?

Giving training is the most urgent and basic activity for any general public. It is necessary to utilize the most recent mechanical advancements to make instructing and learning technique vivid, viable and enjoyable for instructors and students. Smartsheets enhance education methods to educate in a manner that gives an improved learning experience to students. Smartsheets are intuitive. They are associated with a framework and work with a projector. These days, they likewise give students the capacity to take notes, draw, and compose by means of tablets. A few focal points like sorting data, improvements in learning productivity, and better comprehension of ideas have led this to this development becoming a fundamental prerequisite of any smart classroom.

With intuitive whiteboards, or smart sheets, presentations are significantly improved. When an image or document is shown on the board, educators can compose on them with an Internet-related stylus that gives a trove of additional information to the exercise, for instance, definitions, extra pictures or showing a video. Educators are additionally ready to file and offer any exercise that has utilized the board, and past exercises can be returned to fortify new subjects being secured.

- Tablets and eBooks: Have helped students in acquiring material more quickly. Less expensive than paper course books, refreshed learning material is given to students. The decreased need for paper is one of the benefits.
- Personalized learning: Collecting information with reference to student performance for the duration of exercises empowers instructors to customize additional learning according to their enthusiasm and abilities. This is increasingly pivotal for elevated investigations where students have improved possibilities at finishing their examinations if the course is customized for them. Associated frameworks identify interruption and students' lack of engagement right away. The students could then be given more engaging material to motivate them to keep learning.
- Student following framework: With smart mechanized frameworks set up, for students it probably won't be important to be available at a particular time at a particular place to be stamped as present.
- Aiding exceptional need students: IoT can be of great help for the extraordinary needs of the students. Students are allowed to learn at their own pace, and their performance improves without undue worry about the remainder of the class. If beneficial, a student can change over to voice content during tests while a vision impeded student can utilize VR headsets to see the smartboard. IoT likewise may decrease demand to set up discrete educational program and instructive offices for students with unique needs.
- Enhanced security for learning: IoT can likewise help with making a protected and secure condition for training. CCTV cameras are utilized for checking school and college spaces and guarding students. On the off chance that they are associated and continuous observing is carried out, prompt action can be made in the event of crises. It has been observed and demonstrated on numerous occasions that a safe instructive condition improves participation, focus, commitment, and learning in the student.
- Smart classrooms: A smart classroom has AR-empowered smartboards that show enhanced images and recordings on subjects being examined. Students use VR headsets to have a vivid experience. A smart classroom has every one of the gadgets associated with the individual devices of students. The educator can consequently stack assignments and activities to be done on their students' gadgets as they show advancement.
- Task-based learning: Class learning is more assignment based in order to guarantee the perception of the theme being studied. Gadgets associated with IoT guarantee that students work together and share information to finish the allocated assignments. IoT gadgets additionally guarantee that students can receive the assistance of the educator at whatever point and any place required.
- AI-fueled research: Scientists have consistently teamed up to accelerate development by learning from one another. Artificial intelligence and machine learning are utilized in order to associate databases, break down research information, and attain knowledge for further exploration.
- Digital tools: Gone are the times of conventional board instruction. Now is the time of smart classrooms, which enable staff to confer training utilizing PowerPoint introductions, word reports, video screening, and audio sessions.

A picture is worth a thousand words! According to this popular adage, in smart classrooms, students will have the option to acclimatize data that is exhibited by means of these instructional tools. Since the educator isn't composing on the board, students will have the option to amass information more readily in class, absorbing various media (AV) data through advanced apparatuses such as pen drives, CDs, and PDF documents being messaged to students.

- Interactive condition: Since learning is connected to photographs, maps, pictures, and recordings, you as a student will have the option to build up a solid interface with staff. Faculties can further share their considerations openly in class, communicating via words and drawings.
- Student ID cards: The most groundbreaking schools and colleges are currently adopting smart identity cards enabling security along with efficiency. A smart card can enable a student to give recognizable proof, make purchases, to permit secure entry, dispense student records, in order to continuously let staff know in which study hall or lobby students are working,. This information provides a report to aid security.
- Temperature and natural sensors: Provides solid learning situations to students.
- Automated participation following frameworks: In the present aggressive environment, with expanding hours for working and less time in the classroom, instructors require equipment to assist them in using time in class proficiently. Rather than concentrating on educating, educators are frequently tasked with tasks such as gauging participation. Physically gauging participation and stamping it in registers takes a lot of time.

To overcome such wasteful procedures, there are different programs available to accelerate the participation procedure and decrease manual work. Online participation or e-participation is one of them, which is created to mechanize the day-by-day participation in schools. Moreover, it keeps precise records and produces condensed student participation reports.

- Wireless entryway secures and lock: Keyless locks are an answer for protecting homeroom entryways. These locks are picking up acceptance on a wide scale. Keyless study hall locks empower quicker securing of study hall entryways, deferring and conceivably keeping a shooter from going into the room.
- Video conferencing: Education is developing quicker than at any other time. Teachers are confronting expanding requests because of new educational plan norms. Advanced education foundations are progressively developing remote learning programs. Utilizing video conferencing, specialists can be used in order to improve learning. Remote learning projects can be enhanced with video giving progressively intuitive and synchronous learning. Staff can effectively teach over video regardless of location.
- Document cameras: These are affordable gadgets that can help to effectively draw in students. Cameras enable teachers to show lecture-related material on a screen. Exercises and tests can be shown with the camera, in order to help study halls become more inviting.

- Learning management: Envision a situation wherein there was an item that could help in specifying and movement of informative courses, coordinating activities, or learning and progression programs, making examinations, evaluating, and study corridor participation. Imparting instruction is the most essential and basic activity for any general public. It is required to utilize most recent mechanical developments to make instructing and learning techniques powerful, viable, and enjoyable for educators and students.

IoT has tremendous potential in the training market, and it is possible to make IoT part of standard education. The primary variables vital in enabling this to happen are:

- availability
- convenience
- less expense

With this massive development, instructive practices have ecological and social expenses. Huge amounts of smart gadgets are available these days. The generation and transfer of these gadgets incorporate:

- possibility of water contamination, air pollution and land toxins
- utilization of an enormous amount of energy
- producing high volumes of waste
- contribution of risky and exploitative exercises.

Thus there is a need to give answers for all previously mentioned issues of advances utilized in the education sector. Society is searching for green solutions. G-IoT (i.e. Green Internet of Things) not only maximizes benefits but also limits the harm to individuals and the environment.

18.2.5 APPLICATIONS

IoT can be used in various ways, for example, appliance improvement, research purposes, network management, visualization, and systematic data management.

1. Multimedia for simple controls.
2. Home mechanization, components management, and security execution.
3. Tracking the logistic assistance.
4. The scheduling aspects.
5. To monitor transportation units.
6. Wellbeing business.
7. E-health framework, such as pulse checking and remote medical procedures.
8. Environmental checking including air, water quality, soil, untamed life observing.
9. Infrastructure and observing urban and provincial resources.

10. Smart parking, shrewd traffic control, vehicle to vehicle communication and so forth.
11. Industrial activities in the nourishment business, farming, surveillance and so on.

18.3 RELATED WORK

Specialists are taking a shot at the infrastructural changes as well as attempting to redesign the educating and learning strategy. E-learning is a model from those new systems. Giovanni et al. brought up some key ideal models to construct advanced e-learning. It can improve conventional learning systems and can be utilized to make another learning model. Analysts likewise sifted through all strategies and methods to get a distinctive learning objective [4]. Jeannette and Vic proposed another model called iLearning to move students to learn future leading edge innovations in IoT empowered facilities called Living Labs and utilized IoT as a stage for showing a software engineering course [5]. Scientists likewise gather information from a recently presented stage for additional innovative work. Hussein and Haifa proposed a model for interconnection, coordinated effort, and persistent investigation of instructive information sent with using IoT. Their proposal gathers information and dissects with an insightful information mining strategy to construct the e-observing and e-assessment framework for scholarly organizations with quality and security [6]. Yazid et al. proposed another IoT keen gadget that is utilized as a halfway gadget among client and cloud administrations. This gadget gives access to e-learning content sharing [7]. In this way, we can without much of a stretch notify late redesigns in the instruction area with the assistance of IoT gadgets. In our work, we center more around intuitive substance advancement which can be utilized for any e-learning condition and is advantageous for students.

18.4 GREEN – IOT

IoT is a worldwide, imperceptible, encompassing correspondence system and regis-tering condition dependent on cameras, keen sensors, databases, programming, and server farms in a world-traversing data texture framework [8]. The examination in [9] embraced the possibility of IoT focused on saving energy. In spite of earlier proof given in [9], IoT components were talked about in [10], where the benefits of IoT engineering with respect to making it green using trend-setting innovations shrewdly and effectively were portrayed.

Green IoT centers around reducing IoT energy use and decreasing CO_2 emis-sions. Green IoT comprises planning and utilizing perspectives. As illustrated in Figure 18.2, components of green IoT allude to creating gadgets, correspond-ence conventions, energy proficiency, and systems administration structures [11]. Utilizing IoT components can lessen or take out CO_2 emissions, decrease the con-tamination, and improve energy proficiency. In spite of earlier proof Uddin et al. [12] displayed the methods for improving energy productivity and decreasing CO_2 for empowering green data innovation. Since M2M are outfitted with sensors and correspondence additional items, they can speak with one another and sense the world. In systems administration, green IoT plans to recognize the area of the

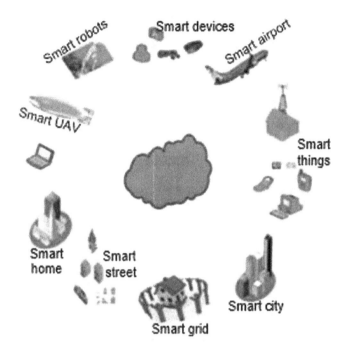

FIGURE 18.2 Green IoT

transfer and number of hubs which fulfill energy sparing and spending requirements. Green IoT has an invaluable job in sending decreasing energy utilization [13], CO_2 emission [14] and contamination [15–17], ecological protection [18], and limiting force utilization [14]. Murugesan additionally characterized green IoT [19] as "the examination and practice of planning, utilizing, assembling, and discarding servers, PCs, and related subsystems, for example, screens, storage gadgets, printers, and correspondence arrange frameworks productively and adequately with negligible or no effect on the earth." Green IoT has three aims, in particular, structure innovations, influence advancements, and empowering advances. Structure advancements allude to the energy effectiveness of gadgets, interchanges conventions, organize models, and interconnections. Influence advancements allude to cutting carbon outflows and improving energy productivity. Because of green ICT innovations, green IoT is progressively proficient in reducing energy, decreasing dangerous emission, lessening assets utilization, and decreasing contamination. Thus, Green IoT prompts protecting common assets, enhances wellbeing, and decreases expenses. Along these lines, green IoT is concentrated on green assembling, green usage, green structure, and green transfer [20].

The world is becoming smarter with the rapid development of science and technology. IoT is a global computing environment that is based on enabling technologies like RFID sensors, cameras, biometrics, software, and data centers. IoT is adopted for the green environment along with minimizing use of energy and time consumption. It not only supports innovations but also addresses various societal challenges.

Major paths addressing computing effects on the environment are:

(i)   Green ambient intelligence
(ii) Green application of IoT
(iii) Green design
(iv) Green creation/manufacturing
(v)   Green scrapping
(vi) Green IoT environment

Green IoT primarily focuses on energy efficiency. Green services in IoT enable a greener society as energy consumption of IoT systems is reduced. Green Internet of Things (G-IoT) was basically introduced to provide significant changes in our daily life. It is the aim of G-IoT to focus on green net generation through IoT enabled technologies for:

• Reducing emission of CO_2 and pollution
• Less operational cost
• Less power/energy consumption
• Capitalize on environmental conservation and surveillance

Few green innovations should be incorporated for the advancements towards green IoT, such as, green Radio Frequency Identification (RFID) labels, green detecting systems, and green distributed computing system,. Figure 18.3 demonstrates advancements to attain a framework for green IoT. RFID is a little electronic device that joins a couple of RFID names and has little mark for every client. RFID marks can store information for the things with which they are associated. The transmission extent of RFID structures is two or three meters. There are two sorts of RFID marks named as unique names and segregated names. The dynamic marks have batteries to continually transmit their own sign while the reserved names don't have their own battery. As opposed to an introduced battery, the segregated marks need to gather energy from the per client signal. To achieve the target of green RFID, a couple of research attempts have been made. One of the proposed game plans is to diminish the size of RFID marks and, right now, the proportion of non-degradable material.

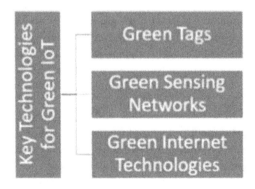

FIGURE 18.3   Key technologies for green IoT

Similarly, various conventions to bring about RFID labels are proposed. In development, different methods are proposed to achieve energy effective RFID names. Also, a green remote sensor position (WSN) is another key development to enable green IoT. Remote sensor mastermind (WSN) contains a huge number of sensor centers with required power and limits. To achieve green WSN, different techniques should be considered:

- The sensor uses energy for the important activity and a while later is placed into inactive or rest mode.
- Use feasible power source for charging and use purposes. Additionally, dynamic energy and vibrations can be used.
- Use energy efficient advancement strategies.
- Use data and setting care figures to reduce the data estimate.
- Use energy effective controlling frameworks to diminish the versatility power usage.

For green web development, hardware, and programming should be considered where gear game plans create devices that use less energy without a decrease of the display. On the other hand, productive plans are offered that use less energy by reduced utilization of the benefits. Also, power saving virtual machine strategies should be realized. As illustrated in Figure 18.3, with respect to green IoT advancement, there are a lot of uses and organizations. It contains splendid urban zones, smart energy and sharp lattice systems, smart establishment, clever assembling plant, and splendid remedial structures.

Within a few years users will be able to manage daily tasks with green support. Applications of G-IoT:

1. Smart education
2. Smart home
3. Medical/healthcare
4. Industrial automation
5. Smart grid
6. Smart city
7. Smart transport
8. Smart manufacturing
9. Smart surveillance
10. Smart logistics and retail
11. Agriculture
12. Telecommunication

Challenges of G-IoT

1. Reliability
2. Complexity reduction
3. Efficient security mechanism
4. Decreasing CO_2
5. Saving energy

FIGURE 18.4   Green IoT applications

18.5   GREEN EDUCATION SYSTEM

Green- IoT was coined for the development of society by keeping the education sector in mind. The technology plays a very important role in this education sector. Figure 18.5 describes the latest trends being followed in smart classrooms. As of now, there are nine main categories of tools, strategies, and technologies that are benefitting the education sector:

- End user technology
- Internet technology
- Teaching technology
- Assessment technology
- Social media technologies
- Visualization technologies
- Digital strategy
- Enabling technology
- Learning technology

FIGURE 18.5   New trends in smart classrooms

The following assignments have the possibility of changing a conventional instructive condition to green educational environment:

- Buy low-cost sensors, renewable energy sources, efficient hard drives, multi-functional equipment (like one system doing printing, scanning faxing and even working as a copier)
- Enable standby or turn off the systems when not in use especially during holidays or at night.
- Improving educational institutes building design. Designing smart campus that minimize consumption of memory, enhanced security, better learning environment.

Concept is
"I hear and I forget.
I see and I believe.
I do and I understand."
    – Confucius

The new technologies in smart classrooms motivate students, along with several other benefits:

- Reduction in waste generation and maximizing recycling and reuse.
- Increase practice of G-IoT among all the stakeholders of education sector.
- Reducing printing and using substitution methods of printing like sending documents as e-mail an attachment.

- Enable remote access to equipment.
- Use of shared service solutions.
- Use of virtual learning sessions like the web conferencing e-learning approach.
- Eliminating information storage that is never again required and increase the amount of low energy storage utilization.

Following G-IoT practices leads to preservation of natural resources, reduction of the impact of technology on the environment, cost reduction, and more. All the stakeholders will benefit, teachers will improve their teaching activities, and students will improve learning and critical thinking exercises in a protected condition, while the executives, authoritative structures, and administration will accomplish huge monetary investment funds that will fundamentally add to economic advancement of the education sector. Going green with IoT will completely transform the education sector. It is anticipated that the education of tomorrow will be monetarily, socially, and ecologically feasible. In view of the performed study, it very well may be reasoned that the G-IoT will assume a key job in accomplishing this objective.

18.6   FUTURE OF RESEARCH DIRECTIONS

Green IoT will change our future to be more green, extremely high QoS, socially and naturally practical, and financially too. These days, most energizing zones center around greening things, for example, green correspondence and systems administration, green plans, green IoT administrations and applications, energy sparing methodologies, coordinated RFIDs and sensor systems, portability, the collaboration of homogeneous and heterogeneous systems, and smart items. The following areas should be looked into to create ideal and productive answers for greening IoT:

1. There is a requirement for UAV to supplant countless IoT gadgets, particularly in agribusiness, traffic, and observing, which will diminish power utilization and contamination. UAV is a promising innovation that will prompt green IoT with minimal effort and high effectiveness.
2. Transmission information from the sensor to the portable cloud is progressively valuable. The sensor-cloud coordinates the remote sensor system and portable cloud. It is an extremely popular and guaranteed innovation for greening IoT. A green informal organization as an assistance (SNaaS) may examine for energy effectiveness of the framework, administration, and WSN.
3. M2M correspondence assumes a basic job to decrease energy use and unsafe outflows. Smart machines must empower mechanized frameworks.
4. Finding reasonable strategies for upgrading QoS parameters (i.e. bandwidth, deferral, and throughput) will contribute adequately and productively to greening IoT.
5. Greening IoT, will require less energy, searching for new assets, limiting IoT's negative effect on the earth.

6. In order to accomplish energy adjusting to support green communication between IoT gadgets, the radio recurrence energy gather ought to be considered.

7. More research is expected to build up the plan of IoT gadgets that diminishes CO_2 emanation and energy utilization. The most basic errand for keen and green natural life is sparing energy and diminishing the CO_2 discharge.

18.7 CONCLUSION

The huge innovation advancement in the twenty-first century has numerous favorable circumstances. Be that as it may, the development of the innovation requires high energy use and increased CO_2 emissions. In this chapter, we survey and recognize the most basic advances utilized for green IoT. ICT insurgency (i.e. FRID, WSN, M2M, correspondence organize, Internet, DC, and CC) has subjectively increased the ability for greening IoT. In light of the basic elements of ICT innovations, the things around us will become more smart to perform explicit undertakings self-governing, rendering of the new kind of green correspondence among human and things and furthermore among things themselves, where data transfer capacity use is augmented and perilous outflow alleviated, and power utilization is decreased ideally. Future recommendations have been addressed for proficiently and adequately improving the green IoT-based applications. This exploration gives adequate knowledge to anybody who wishes to examine the field of green IoT. The patterns and imminent fate of green IoT are given.

BIBLIOGRAPHY

[1] K. Ashton, That "Internet of Things" thing in the real world, things matter more than ideas, *RFID Journal*, 2009.

[2] D. Brock, The electronic product code (EPC) a naming scheme for physical objects, Auto-ID Center, White Paper, 2001.

[3] International Telecommunication Union, ITU internet report 2005: the internet of things, International Telecommunication Union, Workshop Report, November 2005.

[4] G. Adorni, M. Coccoli, and I. Torre, Semantic web and Internet of Things supporting enhanced learning, *Journal of e-Learning and Knowledge Society*, 8(2) 23–32, 2012.

[5] J. Chin and V. Callaghan, Educational living labs; a novel Internet-of-Things based approach to teaching and research, in 9th IEEE International Conference on Intelligent Environments, Athens, Greece, 2013.

[6] H. Y. AbuMansour and H. Elayyan, IoT theme for smart datamining-based environment to unify distributed learning management systems, in International Conference on Information and Communication Systems (ICICS), Irbid, Jordan, 2018.

[7] M. Yazid Idris, D. Stiawan, N. Mohd Habibullah, A. H. Fikri, M. R. Abd Rahim, and M. Dasuki, IoT smart device for e-learning content sharing on hybrid cloud environment, in EECSI, Yogyakarta, Indonesia, 2017.

[8] V. Doknić, *Internet of Things Greenhouse Monitoring and Automation System*, 2014.

[9] H.-I. Wang, Constructing the green campus within the internet of things architecture, *International Journal of Distributed Sensor Networks*, 10, 804627, 2014.

[10]   J. Gubbi, R. Buyya, S. Marusic, and M. Palaniswami, Internet of Things (IoT): A vision, architectural elements, and future directions, *Future Generation Computer Systems*, 29, 1645–1660, 2013.

[11]   A. Gapchup, A. Wani, A. Wadghule, and S. Jadhav, Emerging trends of green IoT for smart world, *International Journal of Innovative Research in Computer and Communication Engineering*, 5, 2139–2148, 2017.

[12]   M. Uddin, and A.A. Rahman, Energy efficiency and low carbon enabler green IT framework for data centers considering green metrics, *Renewable and Sustainable Energy Reviews*, 16, 4078–4094, 2012.

[13]   IMT Vision-Framework and Overall Objectives of the Future Development of IMT for 2020 and Beyond, document Rec. ITU-R M.20830, 2015.

[14]   F.K. Shaikh, S. Zeadally, and E. Exposito, Enabling technologies for green internet of things, *IEEE Systems Journal*, 11(2), 983–994, 2017.

[15]   C. Xiaojun, L. Xianpeng, and X. Peng, IOT-based air pollution monitoring and forecasting system, in 2015 International Conference on Computer and Computational Sciences (ICCCS), IEEE, 2015, 257–260.

[16]   S. Manna, S.S. Bhunia, and N. Mukherjee, Vehicular pollution monitoring using IoT, in International Conference on Recent Advances and Innovations in Engineering (ICRAIE-2014), IEEE, 2014, 1–5.

[17]   T. Zupancic, C. Westmacott, and M. Bulthuis, *The impact of green space on heat and air pollution in urban communities: A metanarrative systematic review*, David Suzuki Foundation Vancouver, BC, Canada, 2015.

[18]   D. Bandyopadhyay and J. Sen, Internet of things: Applications and challenges in technology and standardization, *Wireless Personal Communications*, 58, 49–69, 2011.

[19]   S. Murugesan and G. Gangadharan, *Harnessing green IT: Principles and practices*, Wiley Publishing, 2012.

[20]   C.S. Nandyala and H.-K. Kim, Green IoT Agriculture and Healthcare Application (GAHA), *International Journal of Smart Home*, 10, 289–300, 2016.

Index

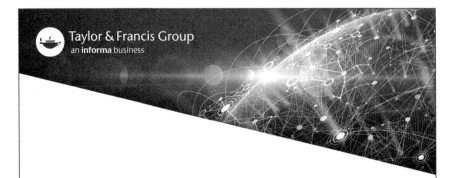

Printed in the United States
by Baker & Taylor Publisher Services